Succeeding in
Applied Calculus

Algebra Essentials

WARREN B. GORDON
Baruch College
The City University of New York

THOMSON

BROOKS/COLE

Australia • Canada • Mexico • Singapore • Spain • United Kingdom • United States

Printed in Canada
1 2 3 4 5 6 7   05 04 03 02 01

Printer: Transcontinental Printing

ISBN: 0-534-40122-8

For more information about our products,
contact us at:
Thomson Learning Academic Resource Center
1-800-423-0563

For permission to use material from this text,
contact us by:
Phone: 1-800-730-2214
Fax: 1-800-731-2215
Web: http://www.thomsonrights.com

Asia
Thomson Learning
5 Shenton Way #01-01
UIC Building
Singapore 068808

Australia
Nelson Thomson Learning
102 Dodds Street
South Street
South Melbourne, Victoria 3205
Australia

Canada
Nelson Thomson Learning
1120 Birchmount Road
Toronto, Ontario M1K 5G4
Canada

Europe/Middle East/South Africa
Thomson Learning
High Holborn House
50/51 Bedford Row
London WC1R 4LR
United Kingdom

Latin America
Thomson Learning
Seneca, 53
Colonia Polanco
11560 Mexico D.F.
Mexico

Spain
Paraninfo Thomson Learning
Calle/Magallanes, 25
28015 Madrid, Spain

TABLE OF CONTENTS

Preface ix

1 Real Numbers and Their Properties 1
Pretest 1
The Real Number System 1
Variables, Algebraic Expressions, Terms, Coefficients and Factors 2
Order of Operations 2
Associative, Commutative, Distributive, Identity, and Inverse Properties 4
Exercises 7
Posttest 9

2 A Review of the Laws of Exponents 11
Pretest 11
Exponents 11
Negative Exponents 12
Sum Rule 12
Product Rule 12
Quotient Rule 13
Zero Rule 14
Distributive Rules 14
Exercises 15
Posttest 17

3 Radicals 19
Pretest 19
Roots 19
Simplification of Radicals 20
Pythagoras' Theorem 23
The Distance Formula 25
Combining Radicals 27
Exercises 28
Posttest 32

4   Multiplication and Rationalization of Radicals   33

Pretest   33
Basic Operations   33
Product of Two Binomials   34
Conjugates   35
Rationalization   36
Indeterminate Forms   39
Exercises   40
Posttest   43

5   Complex Numbers   45

Pretest   45
Imaginary Number *i*   45
Complex Numbers   46
Exercise   48
Posttest   50

6   Fractional Exponents   51

Pretest   51
Fractional Roots   51
Equations with Fractional Exponents   53
Products of Radicals   54
Compound Interest   54
Rule of 72   55
Simple Pendulum   56
Exercises   56
Posttest   59

7   A Review of Factoring   61

Pretest   61
GCF   61
Monomial Factors   64
Factoring Trinomials   65
Perfect Squares   70
Conjugates - The Difference of Two Squares   70
Factoring by Grouping   71
Quadratic Equations and Factoring   72
Exercises   73
Posttest   75

8   Solving Linear Equations 77

Pretest 77
Addition and Multiplication Properties 77
Linear Equations with Fractions 79
Linear Equations with Decimals 80
Solving for a Particular Variable 81
Applications of Linear Equations 82
Exercise 86
Posttest 88

9   Solving Equations of the Form $ax^2 + b = 0$ 89

Pretest 89
Isolation of Squared Term 89
Isolation of Squared Binomial Term 91
Exercises 93
Posttest 94

10   Completing the Square 95

Pretest 95
Completion of the Square 95
Exercise 99
Posttest 101

11   The Quadratic Formula and Applications 103

Pretest 103
Quadratic Formula 104
Clearing Fractions 106
Applications 106
Equations Reducible to Quadratics 107
Exercises 108
Posttest 111

12   Observations and Extensions 113

Pretest 113
Discriminant 113
Sum and Product of Roots 116
Exercises 117
Posttest 119

13   Equations Containing Radicals   121
    Pretest   121
    Isolation and Extraneous Roots   121
    Multiple Square Roots   124
    Exercises   125
    Posttest   127

14   Solving Non-Linear Inequalities   129
    Pretest   129
    Sign Analysis   130
    Interval Notation   131
    Exercises   138
    Posttest   140

15   The Line   141
    Pretest   141
    Two Dimensional Coordinate System   142
    Horizontal and Vertical Lines   143
    The Slope Intercept Form   144
    Graphing   145
    The Point Slope Form   146
    The Slope Formula   148
    The General Linear Equation   151
    An Economic Application   153
    Exercise   156
    Posttest   160

16   The Circle   161
    Pretest   161
    Definition of a Circle   161
    Equation of a Circle   162
    Graphing a Circle   163
    The Ellipse   168
    Exercises   170
    Posttest   171

17   Solving Two Equations in Two Unknowns   173
    Pretest   173
    Method of Elimination   173
    Method of Substitution   176
    Exercises   178
    Posttest   180

18   Non-Linear Systems of Equations   181
Pretest   181
Number of Solutions   181
Method of Substitution   182
Method of Elimination   187
A Calculus Application   191
Exercises   196
Posttest   198

19   Appendix - Arithmetic of Signed Numbers   199
Pretest   199
Absolute Value   199
Addition of Signed Numbers   200
Subtraction of Signed Numbers   201
Multiplication of Signed Numbers   201
Division of Signed Numbers   203
Exercises   203
Posttest   204

20   ANSWERS TO PRE AND POST TESTS   205

21   ANSWERS TO EXERCISES   209

22   INDEX   223

Preface

The reason many students have difficulty with calculus has very little to do with the calculus itself and more to do with a lack of algebra skills required for the course. This may occur for a variety of reasons, the most common of which is the time gap since you have last taken mathematics and your algebraic skills are "rusty." Or, you lack confidence that you really "got it" in the first place. **This book is written so that students in need of an algebra refresher may have a convenient source for reference and review while concurrently taking an Applied Calculus course** –rather than having to "re-take" an algebra course

FOCUS

This book was written with calculus in mind; whenever possible, examples selected are similar to those that will arise in a calculus course. Therefore, you will find this book streamlined to review those algebra skills that are directly applied, and in context, to the concepts of calculus. For example, in reviewing factoring, we consider expressions that would arise in the simplification of a derivative, when reviewing rationalization, we introduce indeterminate forms that arise when working with limits and the definition of the derivative. Section 14 - Non Linear Inequalities, was written with the first and second derivative test in mind, and the last few examples in Section 18 - Non Linear Systems of Equations, illustrate the algebraic technique used in the method of Lagrange multipliers.

You will note that some topics are missing from this text, in particular discussions of functions, inverses, parabolas, exponential, logarithmic and trigonometric functions. These were omitted since most applied calculus texts review these topics in considerable detail, with the possible exception of trigonometric functions. However, we provide a detailed discussion of these topics, written is the same style as this text, on our website at www.brookscole.com/math/authors/gordon, where you may download these pages should you need them.

JUST-IN-TIME APPROACH

This book is designed to be used, as needed, throughout your Applied Calculus course. The inside front cover of this book illustrates which topics in this book are required for the corresponding concepts in Applied Calculus. Each chapter begins with a **Pretest**, which is designed to help you assess your skills prior to working through the chapter. If you feel that you already know the material in a particular section, take the pretest and see how well you do. The total point value on each Pre and Post Test is ten. If your score is eight or more, you may confidently move on to the next section. Each chapter concludes with a Posttest which, after completing the chapter, should assess your mastery with the concept. Therefore, we recommend that you plan ahead and cover the needed chapters in this book before you attempt to cover the calculus concept in your textbook.

COVERAGE

The student must be aware that reading a text on mathematics (at any level) is different from reading texts in other disciplines. You must read mathematics with pencil and paper; working the problems *along with* the book, being certain that you understand each step in its solution. Proficiency is gained with practice! We have included many exercises so that you may reinforce your understanding by working through them. Moreover, we have included answers to both the odd and even exercises.

Calculators are now an integral part of mathematics. While this has distinct advantages, especially in calculus, most of the exercises in this book should be done without one, with the exception of the sections involving exponential equations where a calculator is necessary.

We assume the student has some familiarity with the more elementary notions from algebra. For example, in our discussion of linear equations, we assume the student recalls how to solve two linear equations in two unknowns using both the elimination and substitution methods; however, a complete review of these methods is discussed in Section 17 so as to provide a natural transition to the solution of non linear systems given in Section 18. We refer to these methods when we solve non-linear systems, where we provide the graphs of the equations so the student may have a visual understanding of the relationship between solutions and intersections of graphs; we do not indicate how these graphs are obtained. On a few occasions, we use function terminology and notation, a review of which may be found on our website indicated above.

Straight lines are included in this text as a thorough understanding of them is fundamental to the understanding of most quantitative disciplines, especially calculus and economics. Our approach here is somewhat unorthodox, as we obtain the formula for the slope of a line *after* obtaining its equation. This makes perfectly good sense, especially for a student who may have seen these concepts before.

ACKNOWLEDGMENTS

I must acknowledge many who have provided me with assistance and advice. First, my friend and colleague Hal Shane from whom I have learned so much through our collaborations over the years; to the reviewers who made many useful comments and suggestions: Karabi Datta, Northern Illinois University, Harvey Lambert, University of Nevada - Reno, Carol Nessmith, Georgia Southern University, Maijian Qian, California State University -Fullerton , and George Yaney, University of South Florida. Special thanks to Ann Day at BrooksCole for guiding my notes into this text and to Curt Hinrichs who recognized a national need, and suggested expanding my notes into this text. I am indebted to my students who have suffered through many typographical errors in preliminary editions of this text; it is from their questions and facial expressions that I have learned so much. There is no way I can acknowledge, in words, the support my wife Barbara has provided, to her I say "shut the door!"

Warren B. Gordon

January, 2002

1. Real Numbers and their Properties

» **The Real Number System**
» **Variables, Algebraic Expressions, Terms, Coefficients and Factors**
» **Order of Operations**
» **Associative, Commutative, Distributive, Identity, and Inverse Properties**

PRETEST 1 - Time 10 minutes

Each question is worth one point.

1. Which of the following is /are irrational number(s)? (a) the number whose square is 9 (b) the number whose square is 225 (c) $\pi/5$ (d) 1.111111...

Simply the expression given in questions 2 - 7.

2. $12 + 3 \cdot 2$ 3. $2 \cdot 3^2$ 4. -2^4

5. Simplify $147 - 3\{2 - 3(4 + 5)\}$ 6. Simplify $2 + 3^2\{60 - 2(3 + 5^2)\}$

7. $15x^2y^7 - 8x^2y^7$

In questions 8 - 10 which property of real numbers justifies the given statement?

8. $5 + (7 + x) = (5 + 7) + x$

9. $3(x + y) = 3x + 3y$

10. $5 + 2x = 2x + 5$

We begin with a brief history of the real number system. By the set of natural numbers we mean the ordinary counting numbers,

$$N = \{1, 2, 3, 4, 5, \ldots\}$$

The Real Number System

This set is of numbers is also called the set of positive integers. These were the first numbers used by ancient man, who eventually enlarged this set with the addition of the number zero; much later came the negative integers. Thus, the set of all integers is

$$Z = \{\ldots -3, -2, -1, 0, 1, 2, 3, \ldots\}$$

This set was then enlarged to include fractions, numbers in the form of a/b where $b \neq 0$. a is called the numerator and b the denominator. Thus, 2/3, -5/7, 21/4, and 7/1 are examples of fractions. (Note the integers are fractions with denominator 1.) This new enlarged set, the set of all numbers which are the quotient of two integers (with a non-zero denominator) is called the set of *rational numbers*. The real number system was then expanded to include those numbers that are not rational, that is, cannot be expressed as the quotient of two integers. Such numbers are called *irrational numbers*, which we shall examine in greater detail later on in this text. Examples of such two numbers are π and the number whose square is two.

What comes next is the combining of the numbers. We make the assumption that you are familiar with signed arithmetic which deals with the addition, subtraction, multiplication and division of the real numbers. (We include a short review of the operations with signed numbers in the Appendix, and recommend you examine it if you are unsure of the rules.) We shall begin here with the study of algebraic expressions and their properties.

In mathematics, symbols are often used to represent unknown quantities. Often we use the letters x, y or z, but any letter may be used. In fact, we recommend the use of a symbol that somehow represents the unknown. For example, if the unknown is Tom's age, let t represent his age, or if the unknown is the distance between two towns, call it d. The unknown symbol is also called a *variable*. Often, we translate compound statements into mathematical statements, for example if t is Tom's age, the statement *two more than three times Tom's age* may be represented as $2 + 3t$. Note than the symbol $3t$ means three times t (where t is Tom's age). Such an expression is called an *algebraic expression,* the component of the expression, $3t$ is called a *term,* as is the component 2. More formally, an algebraic expression is a collection of numbers and symbols (terms) and arithmetic operations. Examples of algebraic expressions are $5x - 2$, $3xy + 4$, $5/z - 3x^2$. Given the term $3x$, the number 3 is called the coefficient of the term, similarly the term $-7xz$ has -7 as its coefficient and the term $9x^2y$ has 9 as its coefficient, the coefficient is the number in the term multiplying the variable(s). When we write two variables together, like xy it should be understood that it means $x{\cdot}y$. Note that in one of the above examples we used an exponent. We shall examine exponents more fully in the next section but for now, recall that 3^2 means the product $3{\cdot}3$, 4^3 means $4{\cdot}4{\cdot}4$, x^2 means $x{\cdot}x$, and so on. Also, when we write xy each term in the product is called a *factor*. You may recall that factoring an algebraic expression means to write it as a product of two or more factors.

Variables, Algebraic Expressions, Terms, Coefficients and Factors

We need to agree on the way to perform operations with algebraic expressions. For example, consider the expression $3 + 4{\cdot}5$. There are two different possible interpretations; we could first perform the multiplication of 4 by 5 and then add 3, obtaining 23, or we could add 3 to 4 and then multiple by 5 obtaining 35. Which is it? Fortunately, mathematicians have agreed on an order of operations in which there is no ambiguity. When there is a choice between multiplication and addition, we always do multiplication first, thus the expression $3 + 4{\cdot}5 = 3 + 20 = 23$. What do we do when the expression to be evaluated is more complicated, perhaps involving many operations? There is a useful mnemonic that helps us remember the proper order of operations, it is *Please Excuse My Dear Aunt Sally* — *P* stands for parenthesis, *E* for exponentiation, *M* for multiplication, *D* for division, *A* for addition and *S* for subtraction. That means, given an expression involving these operations, we first simplify expressions within parentheses, braces or brackets (or any other grouping symbol), next compute all exponents, next do all multiplications or divisions in the order in which they appear (working

Order of Operations

from left to right) and then all additions or subtractions in the order in which they appear (working from left to right).

In the example $3 + 4 \cdot 5$ (where the dot is used to indicate multiplication) we followed the order of operations by first doing the multiplication yielding $3 + 20$ and then the addition obtaining 23. Sometimes, we use a grouping symbol like a parenthesis or bracket to indicate multiplications. For example, $2(3 + 4)$ means $2(7) = 14$.

Example 1
Compute the value of the expression $3 + 2(5 + 3)$.

Solution
We have, following the order of operations,
$$3 + 2(5 + 3) = 3 + 2(8) = 3 + 16 = 19$$
■■■

It should now be clear that the algebraic expression $3 + 2x$ cannot be simplified any further unless we know the value of x. It would be wrong to write $5x$ since that means you would do addition (adding the 3 and 2) before the multiplication (the 2 and x), a violation of the order of operations. This is an important observation, students sometimes make mistakes like this. However, if we knew that $x = 5$, then $3 + 2x = 3 + 2(5) = 3 + 10 = 13$.

Example 2
Compute the value of the expression $8 + 3^2 - 2(3 - 1)$.

Solution
$$8 + 3^2 - 2(3 - 1) = 8 + 9 - 2(3 - 1) = 8 + 9 - (2) = 17 - 2 = 15$$
■■■

Example 3
Compute the value of the expression $3 + 2\{ -3 + 2(3 + 5)\}$.

Solution
$$3 + 2\{ -3 + 2(3 + 5)\} = 3 + 2\{ -3 + 2(8)\} = 3 + 2\{ -3 + 16\} = 3 + 2\{ 13 \} = 3 + 26 = 29$$
■■■

Note that when an expression contains nested parentheses (parentheses within parentheses), it is best to simplify the inner parenthesis first, working your way back to the outermost one.

Example 4
Compute the value of the expression $\dfrac{5 - 3^2 + 4[-3 + 4(7 - 2)]}{6(5 - 2) + 2(3)}$.

Solution

$$\frac{5-3^2+4[-3+4(7-2)]}{6(5-2)+2(3)} = \frac{5-9+4[-3+4(7-2)]}{6(3)+6} = \frac{5-9+4[-3+4(5)]}{18+6} =$$

$$\frac{5-9+4[-3+20]}{24} = \frac{5-9+4[17]}{24} = \frac{5-9+68}{24} = \frac{64}{24} = \frac{8}{3}$$

∎∎∎

Example 5

What is the difference between -2^2 and $(-2)^2$?

Solution

The order of operations tells us that when we evaluate -2^2 we first do the exponentiation so we have

$$-2^2 = -4$$

However, in computing $(-2)^2$ the negative sign is included in the parenthesis, thus $(-2)^2 = (-2)(-2) = 4$.

∎∎∎

Given any algebraic expression, if we are also given the values of the variables, then by substitution, we may evaluate the value of the algebraic expression, as illustrated in the next example.

Example 6

Compute the value of the algebraic expression $3x + 2xyz^2$ if $x = 2$, $y = 3$ and $z = -4$.

Solution

Substituting,

$$3x + 2xyz^2 = 3(2) + 2(2)(3)(-4)^2 = 3(2) + 2(2)(3)(16) = 6 + 192 = 198$$

∎∎∎

In performing operations on algebraic expressions, we often use certain properties which we shall discuss next. It is through the use of these properties that we are able to effectively solve equations for the unknown quantities.

The associative property states that when we add or multiply numbers, the order in which we perform the operation does not matter. For example, what do we mean when we write $3 + 2 + 5$? The associative property states that we can evaluate the expression either as $(3 + 2) + 5$ or $3 + (2 + 5)$, do you see that in either case we obtain 10 as the result? Similarly, what do we mean by $3 \cdot 4 \cdot 2$? The associative property states that we may evaluate the expression either as $3(4 \cdot 2)$ or $(3 \cdot 4)2$, obtaining the product 24. More formally the associate property states that

Associative, Commutative, Distributive, Identity, and Inverse Properties

$$(A + B) + C = A + (B + C)$$

and

$$(AB)C = A(BC)$$

Note that there are two associative properties, one with respect to addition and the other with respect to multiplication.

Consider the expression $2 - 3x + 5 + 7x$. It is the associative property that allows us to rewrite the expression as $2 + 5 - 3x + 7x = 7 + 4x$. (Where did we apply the associative property?)

The commutative property states that if we add or multiply any two numbers, the order in which we perform the operation does not matter, for example $3 + 4 = 4 + 3$ and $4 \cdot 3 = 3 \cdot 4$. More formally,

$$A + B = B + A$$

and

$$AB = BA$$

Like the associative property, there are two commutative properties, one with respect to addition and the other with respect to multiplication.

The third property is the distributive property. You will see that it is precisely this property that explains how we combine similar terms and why we may factor expressions. Consider the expression

$$5(7 + 3)$$

of course this expression can be computed as $5(10) = 50$. But also note that if we *distribute* the 5, that is, write the expressions as $5(7) + 5(3)$ we obtain $35 + 15 = 50$, the same result. That this is true in general follows from the distributive property, which we write formally as

$$A(B + C) = AB + AC$$

This property generalizes, so for example, if there were three terms within the parenthesis, we have

$$A(B + C + D) = AB + AC + AD$$

Sometimes, we write the distributive property as

$$(B + C)A = BA + CA$$

(This variation follows from the commutative property.)

A most important illustration of the distributive property explains the rule for combining algebraic expressions differing only in their coefficients. For example, $3x + 5x = (3 + 5)x = 8x$. We just used the distributive property in reverse. Similarly, $7x^2y^3 - 2x^2y^3 = (7 - 2)x^2y^3 = 5x^2y^3$. Thus, as a consequence of the distributive property, to add two algebraic expression with the identical variable part, add their coefficients and attach this sum to the variable part. Note that we cannot combine different algebraic expressions, for example $3x + 2x^2$ cannot be combined because their variable parts are different.

It is also important to observe that the distributive property gives us the method we shall use to factor algebraic expressions. Consider

$$(A + B)(C + D)$$

If we let $(A + B) = E$, then we have

$$(A + B)(C + D) = E(C + D) = EC + ED$$

Now substituting for E we have

$$(A + B)(C + D) = (A + B)C + (A + B)D = AC + BC + AD + BD$$

or

$$AC + BC + AD + BD = (A + B)(C + D)$$

Thus, it is the distributive property which justifies why the expression $AC + BC + AD + BD$ may be wrttien as the product of the two factors given above.

Example 7
Which property of the real numbers justifies each of the following?
(a) $7 + (2 + 9) = (7 + 2) + 9$
(b) $x3 = 3x$
(c) $7x^2 - 3x^2 = 4x^2$
(d) $2(x - 7) = 2x - 14$

Solution
(a) This follows from the associative property.
(b) This follows from the commutative property.
(c) By the distributive property $7x^2 - 3x^2 = (7 - 3)x^2 = 4x^2$.
(d) By the distributive property $2(x - 7) = 2(x) - 2(7) = 2x - 14$.

■■■

Use of the distributive property sometimes enables us to do "mental" arithmetic as the next example illustrates.

Example 8
Use the distributive property to find the product of 8 and 53.

Solution
We can write $8(53) = 8(50 + 3) = 8(50) + 8(3) = 400 + 24 = 424$.
■■■

The remaining properties frequently used in algebraic manipulations are the identity and inverse properties.

If a is any real number, we have

$$a + 0 = a$$

and

$$a \cdot 1 = 1$$

Thus, adding zero to a number does not change the number nor does multiplication of a number by 1 change the number, each of these operations preserves (the identity) of a. These two properties are called the identity properties.

The inverse properties state that for any real number a, we have another number $(-a)$ with the property that

$$a + (-a) = 0$$

and for every $a \neq 0$, there exits another number $1/a$ such that

$$a \cdot \frac{1}{a} = 1$$

We note that $-a$ is called the *additive inverse* of a, and $1/a$ is called the *multiplicative inverse* of a.

So -5 is the additive inverse of 5 since $5 + (-5) = 0$, and $2/3$ is the multiplicative inverse of $3/2$ since $\frac{3}{2} \cdot \frac{2}{3} = 1$.

Note that our basic rule for cancellation of fractions follows from the multiplicative identity property, For example,

$$\frac{36}{24} = \frac{3}{2} \cdot \frac{12}{12} = \frac{3}{2} \cdot 1 = \frac{3}{2}$$

Example 9
Which property of the real numbers justifies each of the following?

(a) $\frac{7}{3} + (-\frac{7}{3}) = 0$ (b) $(\frac{7}{3})(\frac{3}{7}) = 1$ (c) $43(1) = 43$ (d) $\frac{50}{125} = \frac{2}{5}$

Solution
(a) This follows from the additive inverse property.
(b) This follows from the multiplicative inverse property.
(c) This follows from the multiplicative identity property.
(d) We have $\frac{50}{125} = \frac{2(25)}{5(25)} = \frac{2}{5} \cdot \frac{25}{25} = \frac{2}{5} \cdot 1 = \frac{2}{5}$. Note the use of the multiplicative identity property.

■■■

Exercise set 1
In exercises 1 - 5 determine the coefficient of the given algebraic expression.

1.　　5x

2.　　-2y

3.　　$6x^2y$

4.　　$21z^3$

5.　　$16x^5 y^7 z^8$

In exercises 6 - 10 determine the factors of the given expression

6.　　8xy

7.　　2xz

8.　　$(x + 2)(x - 3)$

9.　　$5x(x - 5)(x + 9)$

10.　　$19x^3y(z - 11)$

In exercises 11- 20 compute the value of the given expression.

11.　　$3 + 2 \cdot 9$

12.　　$18 - 5 \cdot 7$

8

13. $4 \cdot 3^2$

14. $5 - 2^2$

15. $23 + 2^2(5 + 3^2)$

16. (a) -3^3 (b) $(-3)^3$

17. (a) -2^4 (b) $(-2)^4$

18. $38 - 2\{3 - 2(3 + 2^2)\}$

19. $28 + 7\{5^2 + (3 + 2^2 \cdot 3)^2 - 14^2\}$

20. $\dfrac{4 - 5^2 + 3[2 - (6 - 8)]}{2^2(3 + 5) + 3(4)}$

In exercises 21 - 25 compute the value of the algebraic expression for the given values of the variables.

21. $2xy$ $\qquad x = 2, y = 3$

22. $x^2y - 3xy$ $\qquad x = -2, y = -3$

23. $2x - 3xy^2z$ $\qquad x = 2, y = -2, \ z = 3$

24. $2x\{3x - 2(y + z^2)\} \ x = 2, y = -1 \ z = -2$

25. $3xy - 2x^2y^3$ $\qquad x = 2, y = -2$

Which of the properties of the real numbers justifies the equality given in exercises 26 - 34?

26. $3 + 2 = 2 + 3$

27. $3(4 \cdot 5) = (3 \cdot 4)5$

28. $4 \cdot 3 = 3 \cdot 4$

29. $2(x + 3) = 2x + 6$

30. $3x - 12 = 3(x - 4)$

31. $5x^3 - 2x^3 = 3x^3$

32. $\dfrac{12}{28} = \dfrac{3}{7}$

33. $31(1) = 31$

34. $\dfrac{12}{7} + (-\dfrac{12}{7}) = 0$

35. Use the distributive property to mentally multiply 9 and 22.

36. Find the product of 102 and 201 by using the distributive property (write $102 = 100 + 2$ and $201 = 200 + 1$).

In exercises 37 - 41 translate the given statement into an algebraic expression

37. 3 more than twice some number

38. 5 times a number decreased by by 3

39. 3 less than 5 times a number

40. 6 more than three times John's age

41. Twice the sum of Ellie's age and 4.

42. Show that 1.23 is a rational number.

43. Show 1.3333... where the 3s continue indefinitely is a rational number.

POSTTEST 1- Time 10 minutes

Each question is worth one point.

1. Which of the following is /are irrational number(s)? (a) the number whose square is 12 (b) the number whose square is 36 (c) 2π (d) 3.33333... .

Simply the expression given in questions 2 - 7.

2. $6 + 5 \cdot 3$

3. $3 \cdot 4^2$

4. $-(-2)^4$

5. Simplify $75 - 4\{5 - 2(7 + 3)\}$

6. Simplify $3 + 2^3\{60 + 2(6 - 4^2)\}$

7. $25x^9y^3 - 18x^9y^3$

In questions 8 - 10 which property of real numbers justifies the given statement?

8. $x + (5 + 3) = (x + 5) + 3$

9. $2(x + y + z) = 2x + 2y + 2z$

10. $3 - 4x = -4x + 3$

(Notes)

2. A REVIEW OF THE LAWS OF EXPONENTS

» **Exponents**
» **Negative Exponents**
» **Sum Rule**
» **Product Rule**
» **Quotient Rule**
» **Zero Rule**
» **Distributive Rule**

PRETEST 2- Time 10 minutes

Simplify each of the following; there should be no negative exponents in the answer. Each question is worth one point.

1. $\left(\dfrac{2}{3}\right)^3$

2. 2^{-3}

3. x^4x^5

4. $(a^3)^4$

5. $(3^{-2})^3(3^4)^2$

6. $\dfrac{x^{12}}{x^3}$

7. $3^7 3^{-7}$

8. $\dfrac{2^{-3}}{3^{-2}}$

9. $(a^2b^{-3}c^4)^2(a^{-2}b^2c^3)^{-2}$

10. $\left(\dfrac{2x^2y^{-3}}{4x^{-3}y^4}\right)^3$

What follows is a review of the basic properties of exponents that are needed in calculus as well as in other quantitative courses, including economics and statistics.

Recall that by b^3 we mean $b \cdot b \cdot b$ and more generally, if n is a positive integer, the expression

$$b^n = b \cdot b \cdots b \qquad (1)$$

Exponents

where we have n factors on the right-hand side of the equation. b is called the *base* and n is called the *exponent*. Thus, $2^4 = 2 \cdot 2 \cdot 2 \cdot 2 = 16$, (2 is the base and 4 is the exponent), similarly,

$$\left(\frac{1}{3}\right)^3 = \frac{1}{3} \cdot \frac{1}{3} \cdot \frac{1}{3} = \frac{1}{27},$$

1/3 is the base and 3 is the exponent.

Negative Exponent

With the definition of an exponent, we can easily, by the examination of some examples, discover their simple properties. Before we do that, lets us first generalize the definition to allow negative exponents. We define a negative exponent as follows.

$$b^{-n} = \frac{1}{b^n} \tag{2}$$

where n is a positive integer.

Thus, $4^{-3} = \frac{1}{4^3} = \frac{1}{4} \cdot \frac{1}{4} \cdot \frac{1}{4} = \frac{1}{64}$.

We are now in a position to discover some of the basic laws of exponents. Consider the product of two expressions with the same base and different exponents, for example, $3^5 \cdot 3^3$. We may rewrite this expression as

$$3^5 \cdot 3^3 = 3 \cdot 3 \cdot 3 \cdot 3 \cdot 3 \cdot 3 \cdot 3 \cdot 3 = 3^8$$

(A product of five 3s followed by a product of three 3s.) Thus, we see that

$$3^5 \cdot 3^3 = 3^{5+3}$$

More generally, we have the following rule.

$$b^m \cdot b^n = b^{m+n} \tag{3}$$

Sum Rule

In words, this rule states that when multiplying exponential expressions with the *same* base we need only add their exponents.

The next rule deals with the exponentiation of an expression containing exponents. Consider the expression

$$\left(4^3\right)^2$$

We have

$$\left(4^3\right)^2 = 4^3 \cdot 4^3$$

Product Rule

We now use the previous rule that in multiplying expressions with the same base we add exponents to obtain

$$\left(4^3\right)^2 = 4^3 \cdot 4^3 = 4^{3+3} = 4^{3 \cdot 2} = 4^6$$

More generally, we have

$$\left(b^m\right)^n = b^{m \cdot n} \tag{4}$$

In words, to raise an exponential expression to a power, we multiply the powers.

Example 1
Simplify the expression $(2^{-3})^2 (2^4)^3$.

Solution
$$(2^{-3})^2 (2^4)^3 = 2^{-3 \cdot 2} 2^{4 \cdot 3} = 2^{-6} 2^{12} = 2^{-6 + 12} = 2^6 = 64$$

■ ■ ■

Observe that our rules are valid for both positive and negative exponents.

From our definition of a negative exponent we can immediately obtain the rule for division of exponential expressions with the same base. First observe that

$$b^{-n} = \frac{1}{b^n}$$

may be rewritten as

$$\frac{1}{b^n} = b^{-n}$$

Thus, consider the expression

$$\frac{b^m}{b^n}$$

We rewrite this expression and apply the rules we already have as follows.

Quotient Rule	$\dfrac{b^m}{b^n} = b^m \cdot \dfrac{1}{b^n} = b^m \cdot b^{-n} = b^{m-n}$

Summarizing,

$$\frac{b^m}{b^n} = b^{m-n} \tag{5}$$

In words, when we have a quotient of two expressions to the same base, we subtract their exponents as indicated.

Example 2
Simplify each of the following. The answer should not contain any negative exponents.

(a) $\dfrac{3^7}{3^4}$ (b) $\dfrac{4^3}{4^5}$.

Solution
In each case we use (5)

(a) $\dfrac{3^7}{3^4} = 3^{7-4} = 3^3 = 27.$

(b) $\dfrac{4^3}{4^5} = 4^{3-5} = 4^{-2} = \dfrac{1}{4^2} = \dfrac{1}{16}.$

■■■

Note that in (b) of the previous example we used (2) to change the negative exponent to a positive one.

There is a special case of (5) that should be noted. Suppose $m = n$, (and $b \neq 0$) then the left-and side is now one, since we are dividing an expression by itself, so we have

$$1 = \dfrac{b^m}{b^m} = b^{m-m} = b^0$$

Zero Rule

Thus we have

$$b^0 = 1 \qquad\qquad (6)$$

We have two more rules to review involving products and quotients of exponential expressions. Consider the expression

$$(xy)^3 = xy \cdot xy \cdot xy = x \cdot x \cdot x \cdot y \cdot y \cdot y = x^3 y^3$$

Distributive Rules

Note that we rewrote the expression using the commutative property of multiplication. What we did in this particular case generalizes and we have

$$(ab)^n = a^n b^n \qquad\qquad (7)$$

Now consider

$$\left(\dfrac{a}{b}\right)^n = \left(ab^{-1}\right)^n = a^n (b^{-1})^n = a^n b^{-n} = \dfrac{a^n}{b^n}$$

where $b \neq 0$, thus we have proven

$$\left(\dfrac{a}{b}\right)^n = \dfrac{a^n}{b^n} \qquad\qquad (8)$$

Note that what we do in each case is to distribute the exponents.

The next example illustrates how the above rules are used.

Example 3

Simplify the expression $\left(\dfrac{a^2 b^{-4}}{c^{-2}}\right)^{-3}$

Solution

We have

$$\left(\frac{a^2 b^{-4}}{c^{-2}}\right)^{-3} =$$ Property (8)

$$\frac{\left(a^2 b^{-4}\right)^{-3}}{\left(c^{-2}\right)^{-3}} =$$ Property (7)

$$\frac{\left(a^2\right)^{-3}\left(b^{-4}\right)^{-3}}{\left(c^{-2}\right)^{-3}} =$$ Property (4)

$$\frac{a^{-6} b^{12}}{c^6} =$$ Property (2)

$$\frac{b^{12}}{a^6 c^6}$$

■■■

For convenience, we summarize the Laws of Exponents given in this section in Table 1. Which are valid when the exponent is any integer. In fact, they generalize and are valid when the exponent is any real number. In Section 6, we examine the case when the exponent is any rational number.

$b^n = b \cdot b \cdots b$	$b^{-n} = \dfrac{1}{b^n}$	$b^m \cdot b^n = b^{m+n}$
$\left(b^m\right)^n = b^{m \cdot n}$	$\dfrac{b^m}{b^n} = b^{m-n}$	$b^0 = 1$
$(ab)^n = a^n b^n$	$\left(\dfrac{a}{b}\right)^n = \dfrac{a^n}{b^n}$	

Table 1: Laws of Exponents

Exercise set 2

In exercises 1 - 4, write the expression in the form of b^n.

1. $3 \cdot 3 \cdot 3 \cdot 3 \cdot 3$

2. $\dfrac{1}{2} \dfrac{1}{2} \dfrac{1}{2}$

3. $5 \cdot 5 \cdot 5 \cdot 5 \cdot 5 \cdot 5 \cdot 5$

4. $(-5) \cdot (-5) \cdot (-5) \cdot (-5) \cdot (-5) \cdot (-5) \cdot (-5)$

In exercises 5 - 32, simplify the given expression, there should be no negative exponents in your answer.

5. 4^{-2}

6. 3^{-3}

7. $\dfrac{1}{2^{-2}}$

8. $\dfrac{1}{3^{-3}}$

9. $\dfrac{1}{x^{-4}}$

10. $\dfrac{1}{y^{-6}}$

11. $a^2\, a^3$

12. $\left(\dfrac{(3x^2)^2}{(5x^{-4})^3}\right)^0$

13. $\left(\dfrac{(8x^3)^4}{(2x^4)^8}\right)^0$

14. $\left(\dfrac{2x^3y^{-4}}{6x^2y^3}\right)^2$

15. $\dfrac{(2x^2y^3)^2(4x^3y)^{-3}}{(6x^2y^{-3})^{-2}}$

16. $\dfrac{x^{-1}+y^{-1}}{x^{-1}-y^{-1}}$

17. $b^5\, b^2\, b^{-4}$

18. $2^5\, 2^4\, 2^{-6}$

19. $3^5\, 3^{-7}\, 3^2$

20. $5^4\, 5^8\, 5^{-9}$

21. $3^{-4}\, 2^7\, 2^{-3}\, 3^7$

22. $2^5\, 3^5\, 3^{-7}\, 2^4\, 2^{-6}\, 3^2$

23. $(3^2)^3$

24. $(2^5)^2$

25. $(2^3)^{-2}$

26. $(x^3)^4$

27. $(x^5)^{-2}$

28. $(3^2\, 2^3)^2$

29. $(4x^2)^3$

30. $(2x^2)^3\,(4x^{-2})^2$

31. $(3y^4)^{-3}\,(2y^{-3})^3$

32. $(4x^{-5})^3\,(8x^{-2})^{-2}$

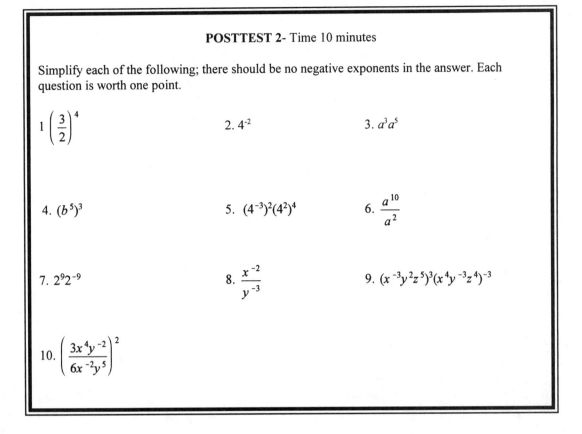

POSTTEST 2- Time 10 minutes

Simplify each of the following; there should be no negative exponents in the answer. Each question is worth one point.

1. $\left(\dfrac{3}{2}\right)^4$

2. 4^{-2}

3. $a^3 a^5$

4. $(b^5)^3$

5. $(4^{-3})^2 (4^2)^4$

6. $\dfrac{a^{10}}{a^2}$

7. $2^9 2^{-9}$

8. $\dfrac{x^{-2}}{y^{-3}}$

9. $(x^{-3} y^2 z^5)^3 (x^4 y^{-3} z^4)^{-3}$

10. $\left(\dfrac{3x^4 y^{-2}}{6x^{-2} y^5}\right)^2$

(Notes)

3. Radicals

» Roots
» Simplification of Radicals
» Pythagoras' Theorem
» The Distance Formula
» Combining Radicals

PRETEST 3 - Time 10 minutes

Each question is worth one point. All answers should be given in simplest form.

Simplify:

1. $\sqrt[3]{27}$

2. $\sqrt[4]{x^{24}}$

3. $2\sqrt{18}$

4. $4\sqrt[3]{16}$

5. $\sqrt{24x^9y^{11}}$

6. $\dfrac{\sqrt{125}}{\sqrt{5}}$

7. $3\sqrt{72} - 4\sqrt{18}$

8. $\sqrt{6}\sqrt{3} + 2\sqrt{50}$

9. If the hypothenuse and one side of a right triangle 12 and 6 are inches respectively, determine the remaining side.

10. Find the distance between the points (2, -2) and (6,6).

You will shortly observe that there is a direct connection between radicals and exponents. A review of exponents was given in the previous section.

We first review a basic notion that you have seen in previous courses, namely the square root. Suppose we ask for the positive number whose square is 4 (of course we know the answer is 2), sometimes we say instead, the square root of 4, written $\sqrt{4}$. Similarly, the square root of 5 would be written as $\sqrt{5}$, and if x is any positive number, the square root of x is written as $\sqrt{x}$.

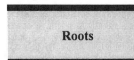

Roots

Similarly, the number whose cube is 8 is written $\sqrt[3]{8}$, and this symbol is read the cube root of eight, the number whose fourth power is 81, the fourth root of 81, is written $\sqrt[4]{81}$, an in general, the number whose nth power is x, the

19

nth root of x, is written $\sqrt[n]{x}$. Note that when $n = 2$, we have the square root, but do not actually write the index 2, it is understood. The backwards-like check symbol used to indicate the nth root, $\sqrt[n]{}$, is called a *radical* symbol, the expression it contains is called the *radicand*, and the entire expression is sometimes called a *radical*.

Any number that may be written in the form of $\frac{m}{n}$ where m and n are integers is said to be *rational*, while any number that can not be written in this form are said to be *irrational*. For example $\sqrt{2}$ is an irrational number because there is no rational number which when multiplied by itself is 2. You can use a calculator to determine an approximate value for $\sqrt{2}$, if you do, you will see that it is approximately 1.414. (The actual proof that $\sqrt{2}$ is irrational is outlined in the exercises.) Similarly, $\sqrt[3]{5}$ is irrational. The square root of any integer that is not a perfect square is irrational, the cube root of any integer that is not a perfect cube is irrational, and so on. However, there are other kinds of irrational numbers that are not related to radicals. For example, the number Pi which is equal to the ratio of the circumference of a any circle to its diameter is an irrational number, and is denoted by the symbol π.

Simplification of Radicals

In this section we will be mostly concerned with the simplification of radicals. Therefore, it is expected that you are familiar with those commonly occurring perfect squares, cubes, fourth, and so on. We list some of them in Table 1.

$\sqrt{4} = 2$	$\sqrt{9} = 3$	$\sqrt{16} = 4$	$\sqrt{25} = 5$
$\sqrt{36} = 6$	$\sqrt{49} = 7$	$\sqrt{64} = 8$	$\sqrt{81} = 9$
$\sqrt{100} = 10$	$\sqrt{121} = 11$	$\sqrt{144} = 12$	$\sqrt{169} = 13$
$\sqrt{196} = 14$	$\sqrt{225} = 15$	$\sqrt{256} = 16$	$\sqrt{289} = 17$
$\sqrt{324} = 18$	$\sqrt{361} = 19$	$\sqrt{400} = 20$	$\sqrt{625} = 25$
$\sqrt{900} = 30$	$\sqrt{1600} = 40$	$\sqrt{2500} = 50$	$\sqrt{3600} = 60$
$\sqrt[3]{8} = 2$	$\sqrt[3]{27} = 3$	$\sqrt[3]{64} = 4$	$\sqrt[3]{125} = 5$
$\sqrt[4]{16} = 2$	$\sqrt[4]{81} = 3$	$\sqrt[4]{256} = 4$	$\sqrt[4]{625} = 5$
$\sqrt[5]{32} = 2$	$\sqrt[5]{243} = 3$	$\sqrt[6]{64} = 2$	$\sqrt[7]{128} = 2$

Table 1 - Frequently Occurring Roots

We remark that if we just needed the approximate value of a radical for some numerical calculation then no simplification is required. We need only enter the radical with a calculator and obtain the needed approximation. However, there are times when we do need the simplification, and it is for such times that the techniques in this and succeeding sections are useful.

Example 1

Evaluate each of the following: (a) $\sqrt{25}$ (b) $\sqrt{\dfrac{4}{9}}$ (c) $\sqrt{16}\sqrt{36}$ (d) $\dfrac{\sqrt[3]{8}}{\sqrt[3]{27}}$ (e) $\sqrt[3]{a^6}$

Solution

(a) $\sqrt{25} = 5$ (b) since $(2/3)^2 = 4/9$, we have $\sqrt{\dfrac{4}{9}} = 2/3$ (c) $\sqrt{16}\sqrt{36} = (4)(6) = 24$ (d) $\dfrac{\sqrt[3]{8}}{\sqrt[3]{27}} = 2/3$,

(f) since $(a^2)^3 = a^6$, we have that $\sqrt[3]{a^6} = a^2$.

■■■

Remark: The radicand for a square root must always be non-negative (in fact, this is true for any even root). This follows immediately from the observation that any real number multiplied by itself an even number of times must be non-negative. For example, it is impossible to multiply a number by itself and obtain -4 (why?) therefore, $\sqrt{-4}$ cannot be a real number. For that reason $\sqrt{x^2} = |x|$. Note that if x were -4 the answer must be 4, that is, |-4|.

Our main concern in this section is the simplification of radicals. Sometimes that will mean rewriting a given radical so that the simplified expression involves the smallest possible radicand. The main result that is most useful in this regard is the following property.

$$\sqrt[n]{ab} = \sqrt[n]{a}\,\sqrt[n]{b} \tag{1}$$

This result may be used as follows: consider

$$\sqrt{8} = \sqrt{4{\cdot}2} = \sqrt{4}\sqrt{2} = 2\sqrt{2}.$$

Our thinking was to determine the largest perfect square which is a factor of 8 and then factor 8 using this perfect square as one of its factors. Since the factor is a perfect square, its square root is a rational number.

Example 2

Simplify $\sqrt{72}$.

Solution

The *largest* perfect square that is a factor of 72 is 36. Therefore, we have

■■■
$$\sqrt{72} = \sqrt{36{\cdot}2} = \sqrt{36}\,\sqrt{2} = 6\sqrt{2}.$$

In the previous example, you might be wondering why we did not write 72 as 9 times 8. 9 is also a perfect square, but it is not the *largest* perfect square which factors 72. We could have used it anyway, but the simplification then takes a little longer, as

$$\sqrt{72} = \sqrt{9{\cdot}8} = \sqrt{9}\sqrt{8} = 3\sqrt{8} = 3\sqrt{4{\cdot}2} = 3\sqrt{4}\sqrt{2} = 3{\cdot}2\sqrt{2} = 6\sqrt{2}.$$

The same simplification technique is used on any nth root, as we illustrate in the next example.

Example 3

Simplify $\sqrt[3]{108}$.

Solution

We seek the *largest* factor of 108 that is a perfect cube. The perfect cubes that are smaller than 108 are 8, 27, and 64. The one that is a factor of 108 is 27. Therefore, we have

$$\sqrt[3]{108} = \sqrt[3]{27 \cdot 4} = \sqrt[3]{27}\sqrt[3]{4} = 3\sqrt[3]{4}.$$

■■■

Sometimes, we may use equation (1) in the reverse direction to simplify radicals. Consider the next example.

Example 4

Simplify $\sqrt{10}\sqrt{6}$.

Solution

$$\sqrt{10}\sqrt{6} = \sqrt{10 \cdot 6} = \sqrt{60} = = \sqrt{4 \cdot 15} = 2\sqrt{15}.$$

■■■

Alternately, in the previous example, we could have written

$$\sqrt{10}\sqrt{6} = \sqrt{5}\sqrt{2}\sqrt{2}\sqrt{3} = \sqrt{5}\,2\sqrt{3} = 2\sqrt{5}\sqrt{3} = 2\sqrt{15}.$$

Note that $\sqrt{2}\sqrt{2} = (\sqrt{2})^2 = 2$, in fact, for any positive number a, we have

$\sqrt{a}\sqrt{a} = (\sqrt{a})^2 = a$, similarly, $\sqrt[3]{a}\sqrt[3]{a}\sqrt[3]{a} = (\sqrt[3]{a})^3 = a$, and in general,

$$\left(\sqrt[n]{a}\right)^n = a \tag{2}$$

In (2) we assume that if n is an even integer, then a is non-negative.

Example 5

Simplify (a) $\sqrt{12x^9y^8}$ (b) $\sqrt[3]{54x^6y^{11}}$

Solution

In (a) we extract the largest perfect square, in (b) the largest perfect cube.

(a) $$\sqrt{12x^9y^8} = \sqrt{4x^8y^8}\sqrt{3x} = 2x^4y^4\sqrt{3x}$$

(b) $$\sqrt[3]{54x^6y^{11}} = \sqrt[3]{27x^6y^9}\sqrt[3]{2y^2} = 3x^2y^3\sqrt[3]{2y^2}$$

■■■

We need to point out a common error that students sometimes make. Note that equation (1) says that it is permissible to write the radical of a *product* as the *product* of the radicals. However, if we change the word product to sum in the last statement we get a false statement. Consider the following illustration.

$$\sqrt{16+9} \neq \sqrt{16} + \sqrt{9}$$

Why are they not equal? The expression on the left is equal to $\sqrt{25} = 5$ while the expression on the right is equal to $4 + 3 = 7$. Similarly, we get a false statement for a difference, that is,

$$\sqrt{25 - 16} \neq \sqrt{25} - \sqrt{16}$$

The radical on the left simplifies to 3, whereas the difference on the right is 1. However, division of radicals, like multiplication is permissible, for example

$$\sqrt{\frac{25}{16}} = \frac{\sqrt{25}}{\sqrt{16}} = \frac{5}{4},$$

and more generally, we have

$$\sqrt[n]{\frac{a}{b}} = \frac{\sqrt[n]{a}}{\sqrt[n]{b}} \tag{3}$$

where we assume b is not zero and a and b have the appropriate signs when n is even.

Example 6

Simplify $\dfrac{\sqrt{40}}{\sqrt{5}}$.

Solution

$$\frac{\sqrt{40}}{\sqrt{5}} = \sqrt{\frac{40}{5}} = \sqrt{8} = \sqrt{4}\sqrt{2} = 2\sqrt{2}.$$

■■■

Once again we remark that if a numerical approximation is required, using a calculator on the original expression would yield the same approximation obtained from the simplified expression.

Pythagoras' Theorem

One application of square roots is to right triangles. Recall, a right triangle is one having one of its angles equal to ninety degrees. Figure 1 illustrates a right triangle, with sides a, b, and c, where c is the longest side which is opposite the ninety degree (right) angle, and is called the *hypothenuse*, the other two sides are called *legs* of the triangle.

Given a right triangle, we have the famous theorem of Pythagoras, namely the square of the hypothenuse is equal to the sum of the squares of the two legs. In symbols, we have the following.

24

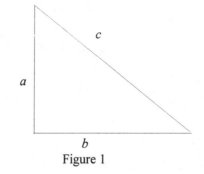

Figure 1

For example, consider the triangle shown in Figure 2. The hypothenuse is determined from the equation

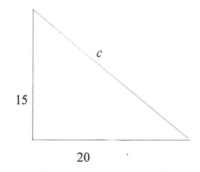

15

20

Figure 2: Illustrating Pythagoras' Theorem

$$c^2 = 15^2 + 20^2$$

or

$$c^2 = 225 + 400 = 625$$

Therefore

$$c = \sqrt{625} = 25$$

Note that -25 is also a solution to the quadratic equation $c^2 = 625$, but c is the length of the hypothenuse, and since length is a positive number, the negative solution is rejected.

There is a very simple proof of Pythagoras' theorem that uses nothing more than the notions than the area of a square and area of a triangle.

Consider the square shown in Figure 3. The length of each side of the external square is $(a + b)$ and the length of each side of the internal square is c. The area of the inside square plus the area of the four internal right triangles is equal to the area of the entire square.

The area of the internal square is c^2. The area of each right triangle is ½ ab, and the area of the external square is $(a + b)^2$. Therefore we have that

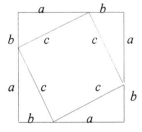

Figure 3: Proving Pythagoras' Theorem

$$(a+b)^2 \; = \; 4 \cdot \frac{1}{2} ab + c^2$$

Multiplying out, we have

$$a^2 + 2ab + b^2 \; = 2ab + c^2$$

or

$$a^2 + b^2 = c^2$$

as required.

∎

Example 7
The hypothenuse and one leg of a right triangle are 13 cm and 12 cm respectively, determine the length of the other leg of the triangle.

Solution
We have $c = 13$ and one leg, say $a = 12$, therefore

$$13^2 = 12^2 + b^2$$

or

$$169 = 144 + b^2$$

or

$$b^2 = 25$$

or

$$b \; = \; \sqrt{25} \; = \; 5\,\text{cm}$$

∎∎∎

One additional remark: it can also be shown that If the sides of any triangle satisfy the Pythagorean theorem, then the triangle is a right triangle.

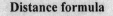
Distance formula

A related application of square roots, which is nothing more than the reformulation of Pythagoras' theorem, is determining the distance between two points in a plane. Suppose we plot the points $A(1,4)$ and $B(6,16)$, what is the distance d between these two points? Draw a horizontal line from $(1,4)$ to the point $C(6,4)$ and a vertical line from $(6,4)$ to $(6,16)$. We now have a right triangle (why?) with the hypothenuse being represented by d, see Figure 4. Note that the horizontal distance of the line segment AC is 6 - 1 = 5, and the vertical distance of the line segment CB is 16 - 4 = 12.

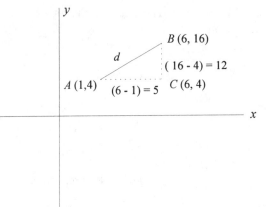

Figure 4: The distance between (1, 4) and (6, 16)

Therefore, by Pythagoras' theorem, we have the distance, d is given by

$$d^2 = 5^2 + 12^2 = 25 + 144 = 169$$

$$d^2 = 169$$

$$d = \sqrt{169} = 13$$

We can generalize this method to find the distance between any two points $A(x_1, y_1)$ and $B(x_2, y_2)$. The required distance is the length of the line segment connecting these two points. As before, we draw a triangle by drawing a horizontal line from A to C (x_2, y_1) and from C to B. The distance of the horizontal line segment AC is $(x_2 - x_1)$, and the distance of the vertical line segment CB is $(y_2 - y_1)$. See Figure 5

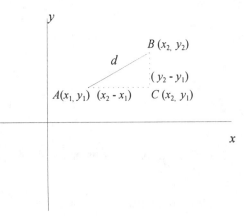

Figure 5: The Distance Formula

Once again, we apply Pythagoras' theorem and we have

$$d^2 = (x_2 - x_1)^2 + (y_2 - y_1)^2$$

or taking the positive square root, we obtain the general distance formula.

The distance d between the points $A(x_1, y_1)$ and $B(x_2, y_2)$ is

$$d = \sqrt{(x_2 - x_1)^2 + (y_2 - y_1)^2}$$

Remark: we plotted the points $A(x_1, y_1)$ and $B(x_2, y_2)$ as if they were in the first quadrant. They could be in any quadrant. In the general case the horizontal and vertical distance are then $|x_2 - x_1|$ and $|y_2 - y_1|$ respectively. Their squares are $(x_2 - x_1)^2$ are $(y_2 - y_1)^2$ respectively, yielding via Pythagoras' theorem, the distance formula

Example 8
Find the distance between the points (1, -3) and (3, 1).

Solution
Let $(x_1, y_1) = (1,-3)$ and $(x_2, y_2) = (3,1)$. Then $(x_2 - x_1) = 3 - 1 = 2$, and $(y_2 - y_1) = 1 - (-3) = 4$. Therefore, we have the distance between these two points d is

$$d = \sqrt{(x_2 - x_1)^2 + (y_2 - y_1)^2} = \sqrt{2^2 + 4^2} = \sqrt{20} = \sqrt{4}\sqrt{5} = 2\sqrt{5}$$

When applying the distance formula, it does not matter which of the points you designate as the first point and which as the second. To verify this remark, rework the preceding example, except now let $(x_1, y_1) = (3,1)$ and $(x_2, y_2) = (1,-3)$.

In Section 1 we reviewed the rules used in the combining of similar terms. For example, $3x + 2x = 5x$, $7xy + 2xy = 9xy$. That is, if the algebraic expressions differ only in their coefficients, we add their coefficients and keep the variable factors the same. The reason follows from the distributive property, for example $3x + 2x = (3 + 2)x = 5x$. Similarly with radicals of the same type, when adding or subtracting them, we add or subtract their coefficients; for example, $3\sqrt{5} + 7\sqrt{5} = 10\sqrt{5}$. The reason is the same as with any algebraic expression, $3\sqrt{5} + 7\sqrt{5} = (3 + 7)\sqrt{5} = 10\sqrt{5}$. Terms that are dissimilar cannot be combined, for example $2\sqrt{3} + 5\sqrt{2}$ cannot be combined. If we need a numerical approximation we would use our calculator.

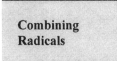
Combining Radicals

Sometimes, radicals *appear* to be different, but in fact, after simplifying they are the same type and can be combined as illustrated in the next example.

Example 9
Simplify $12\sqrt{8} - 2\sqrt{50}$.

Solution

Our objective is to simplify each of the two radicals using the methods discussed earlier in this section.

$$12\sqrt{8} - 2\sqrt{50} = 12\sqrt{4}\sqrt{2} - 2\sqrt{25}\sqrt{2}$$
$$12 \cdot 2\sqrt{2} - 2 \cdot 5\sqrt{2} = 24\sqrt{2} - 10\sqrt{2} =$$
$$14\sqrt{2}$$

■■■

What is true for adding and subtracting square roots is also true for any root. The next example illustrates the same method on a cube and fourth root.

Example 10

Simplify each of the following:

(a) $2\sqrt[3]{81} + 5\sqrt[3]{24}$ (b) $5\sqrt[4]{32x^5} + 7x\sqrt[4]{162x}$

Solution

(a)

$$2\sqrt[3]{81} + 5\sqrt[3]{24} = 2\sqrt[3]{27}\sqrt[3]{3} + 5\sqrt[3]{8}\sqrt[3]{3}$$
$$2 \cdot 3\sqrt[3]{3} + 5 \cdot 2\sqrt[3]{3} = 6\sqrt[3]{3} + 10\sqrt[3]{3} =$$
$$16\sqrt[3]{3}$$

(b)

$$5\sqrt[4]{32x^5} + 7x\sqrt[4]{162x} = 5\sqrt[4]{16x^4}\sqrt[4]{2x} + 7x\sqrt[4]{81}\sqrt[4]{2x} =$$
$$5 \cdot 2x\sqrt[4]{2x} + 7x \cdot 3\sqrt[4]{2x} = 10x\sqrt[4]{2x} + 7x \cdot 3\sqrt[4]{2x} =$$
$$10x\sqrt[4]{2x} + 21x\sqrt[4]{2x} = 31x\sqrt[4]{2x}$$

■■■

Other variations with expressions containing radicals can occur, for example, sometimes we may need to first multiply before simplifying, or perhaps some other operation.

Example 11

Simplify $\sqrt{8}\sqrt{3} + 4\sqrt{54}$.

Solution

$$\sqrt{8}\sqrt{3} + 4\sqrt{54} = \sqrt{24} + 4\sqrt{9}\sqrt{6} =$$
$$\sqrt{4}\sqrt{6} + 4\sqrt{9}\sqrt{6} = 2\sqrt{6} + 4 \cdot 3\sqrt{6} =$$
$$2\sqrt{6} + 12\sqrt{6} = 14\sqrt{6}$$

■■■

Exercise set 3

In Exercises 1-28, Simplify each of the following:

1. $\sqrt{9}$ 2. $-\sqrt{9}$ 3. $-\sqrt{144}$

4. $\sqrt{900}$

5. $\sqrt{\dfrac{25}{16}}$

6. $\sqrt{\dfrac{121}{169}}$

7. $-\sqrt{\dfrac{49}{64}}$

8. $\sqrt{\dfrac{16}{625}}$

9. $\sqrt[3]{27}$

10. $\sqrt[3]{\dfrac{8}{27}}$

11. $\sqrt[3]{-27}$

12. $\sqrt[4]{16}$

13. $\sqrt[4]{\dfrac{81}{16}}$

14. $\sqrt{9+16}$

15. $\sqrt{25-9}$

16. $\sqrt{5^2+12^2}$

17. $\left(\sqrt{7}\right)^2$

18. $\left(\sqrt{21}\right)^2$

19. $\left(\sqrt{17+x}\right)^2$

20. $\left(\sqrt{x^2+9}\right)^2$

21. $\left(\sqrt[3]{3}\right)^3$

22. $\left(\sqrt[3]{4}\right)^3$

23. $\left(\sqrt[3]{-7}\right)^3$

24. $\left(\sqrt[3]{x^3+3^3}\right)^3$

25. $\sqrt{9}+\left(\sqrt{5}\right)^2$

26. $\left(\sqrt{6}\right)^2-2\left(\sqrt{2}\right)^2$

27. $-\sqrt[3]{-8}-\sqrt[3]{8}$

28. $\left(\sqrt[3]{-4}\right)^3-2\left(\sqrt{3}\right)^2$

Approximate, using your calculator to three decimal places in exercises 29- 34.

29. $\sqrt{278}$

30. $\sqrt{0.0739}$

31. $\sqrt{3.78}$

32. $\sqrt[3]{15}$

33. $\sqrt[3]{38}$

34. $\dfrac{\sqrt[3]{425}}{\sqrt{27}}$

In exercises 35 - 61, simplify the given radical expression.

35. $\sqrt{24}$

36. $\sqrt{48}$

37. $\sqrt{32}$

38. $\sqrt{12}$

39. $\sqrt{50}$

40. $\sqrt{80}$

41. $\sqrt{63}$

42. $-\sqrt{120}$

43. $6\sqrt{18}$

44. $3\sqrt{45}$

45. $-5\sqrt{27}$

46. $4\sqrt{56}$

47. $-3\sqrt{300}$

48. $6\sqrt{320}$

49. $-23\sqrt{800}$

50. $\sqrt{6}\sqrt{3}$

51. $\sqrt{7}\sqrt{56}$

52. $\sqrt{12}\sqrt{15}$

53. $\sqrt{12}\sqrt{30}$

54. $3\sqrt{10}\sqrt{20}$

55. $\sqrt{\dfrac{5}{6}}\sqrt{30}$

56. $\sqrt{\dfrac{10}{11}}\sqrt{110}$

57. $\sqrt{\dfrac{5}{8}}\sqrt{\dfrac{5}{2}}$

58. $\sqrt{\frac{12}{15}}\sqrt{300}$

59. $\frac{20\sqrt{15}}{4\sqrt{3}}$

60. $\frac{30\sqrt{80}}{5\sqrt{5}}$

61. $\frac{25\sqrt{120}}{75\sqrt{6}}$

In 62 -74, simplify, assume x, y, and z are non-negative.

62. $\sqrt{x^4}$

63. $\sqrt{y^2}$

64. $\sqrt{z^8}$

65. $\sqrt{x^4 y^6}$

66. $\sqrt{4x^8 z^{10}}$

67. $\sqrt{x^9}$

68. $\sqrt{z^{15}}$

69. $\sqrt{x^9 z^{15}}$

70. $\sqrt{8x^6}$

71. $\sqrt{8x^6 y^7}$

72. $\sqrt{12x^6 y^7}$

73. $\sqrt{50x^5 y^4 z^{11}}$

74. $\sqrt{80x^{25} y^{14} z^{17}}$

In exercises 75-99, simplify the radical expression.

75. $\sqrt[3]{24}$

76. $\sqrt[3]{16}$

77. $\sqrt[3]{54}$

78. $\sqrt[3]{128}$

79. $\sqrt[3]{-128}$

80. $-\sqrt[3]{-81}$

81. $\sqrt[4]{32}$

82. $\sqrt[4]{162}$

83. $\sqrt[3]{24}\cdot\sqrt[3]{9}$

84. $\sqrt[3]{5}\cdot\sqrt[3]{50}$

85. $\sqrt[4]{4}\cdot\sqrt[4]{20}$

86. $\sqrt[3]{x^6}$

87. $\sqrt[3]{x^9}$

88. $\sqrt[4]{y^{12}}$

89. $\sqrt[5]{y^{20}}$

90. $\sqrt[3]{x^7}$

91. $\sqrt[4]{y^{14}}$

92. $\sqrt[3]{y^6 z^{25}}$

93. $\sqrt[3]{8y^6}$

94. $\sqrt[3]{16y^6}$

95. $\sqrt[3]{16y^7}$

96. $\sqrt[3]{16y^7 z^{12}}$

97. $\sqrt[4]{32x^{15} y^{12} z^3}$

98. $\sqrt[4]{162x^{16} y^{23} z^{15}}$

99. $\sqrt[5]{64x^{10} y^{23}}$

Find the unknown side of a right triangle with legs a and b and hypothenuse c. Leave the answer in the simplest radical form.

100. $a = 15$, $b = 20$, $c = ?$

101. $a = 12$, $c = 20$, $b = ?$

102. $b = 15$, $c = 39$, $a = ?$

103. $a = b = 4$, $c = ?$

104. $a = b = 8$, $c = ?$

105. $a = 6$, $b = 12$, $c = ?$

106. $a = b$, $c = 2\sqrt{2}$

107. $a = 2$, $b = 2\sqrt{3}$, $c = ?$

108. Each side of a square is 8 inches. To the nearest thousandth, find the length of its diagonal.

109. Sam leaves Albany and drives north at 30 miles per hour. Jane leaves at the same time and drives east at 20 miles per hour. How far apart are they after 2 hours?

110. A rectangular room has its length equal to three times its width. If the distance from one corner to a diagonally opposite corner is 20 feet, find the dimensions of the room to the nearest one-hundredth of a foot.

In exercises 111- 166, find the distance between the given points. Give the answer both in simplest radical form, as well as to the nearest one-thousandth.

111. $(1, 5), (4,9)$

112. $(7, 9), (12, 21)$

113. $(-5, 3), (2, -4)$

114. $(2, -3), (-3, 5)$

115. $(3/4, 3), (½, -5)$

116. $(1/3,-1/4), (-1/6, 1/8)$

Simplify the given radical expression

117. $4\sqrt{5} + 7\sqrt{5}$

118. $3\sqrt{11} - 5\sqrt{11}$

119. $17\sqrt{21} + 19\sqrt{21}$

120. $3\sqrt[3]{5} + 9\sqrt[3]{9}$

121. $6\sqrt[4]{11} + 5\sqrt[4]{11}$

122. $12\sqrt[5]{7} - 8\sqrt[5]{7}$

123. $3\sqrt{2} - 5\sqrt{8}$

124. $12\sqrt{3} + 4\sqrt{12}$

125. $5\sqrt{8} - 4\sqrt{18}$

126. $6\sqrt{12} - 9\sqrt{48}$

127. $8\sqrt{8} - 3\sqrt{18}$

128. $-21\sqrt{50} - 11\sqrt{32}$

129. $4\sqrt{12} - 2\sqrt{48}$

130. $6\sqrt{20} - 4\sqrt{45}$

131. $2\sqrt{45} - 3\sqrt{80}$

132. $4\sqrt{54} - 5\sqrt{24}$

133. $9\sqrt{28} + 3\sqrt{63}$

134. $9\sqrt{40} + \frac{2}{3}\sqrt{90}$

135. $\frac{2}{3}\sqrt{24} - \frac{3}{4}\sqrt{54}$

136. $\frac{5}{6}\sqrt{45} - \frac{2}{5}\sqrt{20}$

137. $5\sqrt{12} - 4\sqrt{48} + 7\sqrt{75}$

138. $7\sqrt{18} + 12\sqrt{72} - 11\sqrt{50}$

139. $9\sqrt{24} - 8\sqrt{54} - 5\sqrt{150}$

140. $4\sqrt{8} - 4\sqrt{27} + 5\sqrt{32} - 6\sqrt{75}$

141. $5\sqrt{125} + 4\sqrt{28} + 2\sqrt{63} - 6\sqrt{75}$

142. $3\sqrt{44} + 5\sqrt{28} - 2\sqrt{99} - 6\sqrt{175}$

143. $5\sqrt{80} + 7\sqrt{18} - 6\sqrt{20} + 8\sqrt{32}$

144. $6\sqrt[3]{24} + 5\sqrt[3]{81}$

145. $12\sqrt[3]{32} - 8\sqrt[3]{108}$

146. $3\sqrt[4]{32} - 7\sqrt[4]{162}$

147. $6\sqrt[3]{24} + 7\sqrt{12} + 5\sqrt[3]{81} - 5\sqrt{27}$

148. $6\sqrt{18} - 7\sqrt[3]{24} - 7\sqrt{50} + 4\sqrt[3]{81}$

149. $6\sqrt{8x^2} + 2\sqrt{18x^2}$

150. $3\sqrt{27x^3y^2} + 2xy\sqrt{75x}$

151. $4\sqrt{45x^4y^3} - 3xy\sqrt{20x^2y}$

152. $4x^2\sqrt[3]{16x^6y^5} + 5x^3y\sqrt[3]{54x^3y^2}$

153. $5\sqrt[3]{24x^7y^6} - 2x^2y^2\sqrt[3]{375x}$

154. $7x^5\sqrt[4]{32x^5y^8} - 5x^2y\sqrt[4]{162x^9y^4}$

155. Find the distance from the point $(3, 4)$ to the line $3x + 4y = 12$.

156. Find the distance from the point (2,-5) to the line $5x + 12y = 15$

157. To prove that $\sqrt{2}$ is irrational, assume the contrary i.e., assume it is rational and can be written in the form m/n, here the integers m and n have no factors in common. Then we have,

$$\sqrt{2} = \frac{m}{n}$$

$$2 = \frac{m^2}{n^2}$$

$$m^2 = 2n^2$$

Therefore, m is even(why?), and may be written in the form $m = 2k$. Thus, we have

$$m^2 = (2k)^2 = 4k^2 = 2n^2$$
$$n^2 = 2k^2$$

Therefore, n is even (why?) and may be written in the form $n = 2l$. Thus, both m and n have a factor in common, a contradiction. Thus, the only assumption made was that $\sqrt{2}$ was rational, that must be incorrect, it is therefore irrational.

POSTTEST 3- Time 10 minutes

Each question is worth one point. All answers should be given in simplest form.

Simplify:

1. $\sqrt[4]{32}$ 　　　　 2. $\sqrt[6]{y^{36}}$ 　　　　 3. $3\sqrt{27}$ 　　　　 4. $5\sqrt[4]{32}$

5. $\sqrt[3]{54x^{12}y^{10}}$ 　　 6. $\dfrac{\sqrt{75}}{\sqrt{3}}$ 　　 7. $3\sqrt{45} - 2\sqrt{54}$ 　　 8. $4\sqrt{5}\sqrt{10} - 3\sqrt{18}$

9. If the hypothenuse and one side of a right triangle are 10 and 6 inches respectively, determine the remaining side.

10. Find the distance between the points (-4, 5) and (5, 14).

4. Multiplication and Rationalization of Radicals

» **Basic Operations**
» **Product of Two Binomials**
» **Conjugates**
» **Rationalization**
» **Indeterminate Forms**

PRETEST 4- Time 15 minutes

Each question is worth 1 point.

Simplify each of the following expressions.

1. $2(2+3\sqrt{5})-3(-4+8\sqrt{5})$

2. $2\sqrt{3}(3-2\sqrt{6})$

3. $(4+3\sqrt{2})(5-2\sqrt{2})$

4. $(3\sqrt{2}-2\sqrt{3})^2$

5. $(3\sqrt{5}-2\sqrt{3})(3\sqrt{5}+2\sqrt{3})$

6. Factor $x^2 - 12$

In each of the following, rationalize the denominator:

7. $\dfrac{12}{\sqrt{8}}$

8. $\dfrac{\sqrt{8}\sqrt{3}}{\sqrt{12}\sqrt{6}}$

9. $\dfrac{10}{\sqrt[3]{4}}$

10. $\dfrac{26}{5-2\sqrt{3}}$

Basic Operations

Multiplication of radicals is nothing more than the repeated use of the distributive property, so there will really be will nothing new with this operation. However, the application of multiplication is useful in modifying the way quotients containing radical expressions appear, and that observation is the main result of this section. The basic ideas of this section are best examined by a sequence of examples.

Example 1
Simplify $(3-2\sqrt{5})+(-6+7\sqrt{5})$

Solution

$$(3 - 2\sqrt{5}) + (-6 + 7\sqrt{5}) = 3 - 6 - 2\sqrt{5} + 7\sqrt{5} = -3 + 5\sqrt{5}$$

■■■

Example 2
Simplify $(3 - 2\sqrt{5}) - (-6 + 7\sqrt{5})$

Solution

$$(3 - 2\sqrt{5}) - (-6 + 7\sqrt{5}) = 3 - 2\sqrt{5} + 6 - 7\sqrt{5} = 9 - 9\sqrt{5}$$

■■■

Example 3
Simplify $4(5 - 7\sqrt{11})$

Solution

$$4(5 - 7\sqrt{11}) = 4 \cdot 5 - 4 \cdot 7\sqrt{11} = 20 - 28\sqrt{11}$$

■■■

Example 4
Simplify $\sqrt{2}(5 - 3\sqrt{10})$

Solution

$$\sqrt{2}(5 - 3\sqrt{10}) = 5 \cdot \sqrt{2} - 3\sqrt{2}\sqrt{10} = 5\sqrt{2} - 3\sqrt{20} =$$
$$5\sqrt{2} - 3\sqrt{4}\sqrt{5} = 5\sqrt{2} - 3 \cdot 2\sqrt{5} = 5\sqrt{2} - 6\sqrt{5}$$

■■■

Example 5
Simplify $3\sqrt{2}(4 + 2\sqrt{10})$

Solution

$$3\sqrt{2}(4 + 2\sqrt{10}) = 12\sqrt{2} + 6\sqrt{2}\sqrt{10} = 12\sqrt{2} + 6\sqrt{20} =$$
$$12\sqrt{2} + 6\sqrt{4}\sqrt{5} = 12\sqrt{2} + 6 \cdot 2\sqrt{5} = 12\sqrt{2} + 12\sqrt{5}$$

■■■

Product of Two Binomials

The next example asks to simplify the product of two binomial expressions involving radicals. The simplification is performed in exactly the same way we would multiply any product of binomials, namely by the application of the distributive rule.

Example 6
Simplify $(4 + 3\sqrt{5})(7 - 2\sqrt{5})$

Solution

$$(4 + 3\sqrt{5})(7 - 2\sqrt{5}) = 4 \cdot 7 + 4 \cdot (-2)\sqrt{5} + 7 \cdot 3\sqrt{5} + (3\sqrt{5})(-2\sqrt{5}) =$$
$$28 + 21\sqrt{5} - 8\sqrt{5} - 6 \cdot 5 = -2 + 13\sqrt{5}$$

■■■

Note that the method of applying the distributive rule for two binomials is sometimes called the *FOIL* method. *F* - first, *O* - outer, *I* - inner, and *L* - last. In the previous example, the identification is

$$\underset{}{(4 + 3\sqrt{5})(7 - 2\sqrt{5})} = \underset{F}{4 \cdot 7} + \underset{O}{4 \cdot (-2)\sqrt{5}} + \underset{I}{7 \cdot 3\sqrt{5}} + \underset{L}{(3\sqrt{5})(-2\sqrt{5})}$$

where *F* is the product of the first term in each binomial, *O* is the product of the outer term from each binomial, *I* is the product of the inner terms, and *L* is the product of the last terms.

Example 7

Simplify $(3 + 5\sqrt{7})^2$

Solution

$$(3 + 5\sqrt{7})^2 = (3 + 5\sqrt{7})(3 + 5\sqrt{7}) = 3 \cdot 3 + 3 \cdot 5\sqrt{7} + 3 \cdot 5\sqrt{7} + 5\sqrt{7} \cdot 5\sqrt{7} =$$
$$9 + 30\sqrt{7} + 25 \cdot 7 = 184 + 30\sqrt{7}$$

■■■

Example 8

Simplify $(3 + 5\sqrt{7})(3 - 5\sqrt{7})$

Solution

$$(3 + 5\sqrt{7})(3 - 5\sqrt{7}) = 9 + 15\sqrt{7} - 15\sqrt{7} - 25 \cdot 7 = -166$$

■■■

Note that in the last example, the inner and outer radical terms were equal and opposite and therefore canceled. More generally,

$$(a + b\sqrt{c})(a - b\sqrt{c}) = a^2 - b^2 c$$

Two such binomials, which are identical except for their connecting sign, are called *conjugates*. The product of conjugates of the above form is *always* a rational number. This observation will be used below to simplify denominators involving sums and differences of square roots.

It should be noted, that we can use the above observation to factor expressions like $ax^2 - b$ when either a or b is not a square. For example, $x^2 - 5 = (x - \sqrt{5})(x + \sqrt{5})$, more generally, we have

Conjugates

$$ax^2 - b = (x\sqrt{a} - \sqrt{b})(x\sqrt{a} + \sqrt{b})$$

Multiplication of the conjugates will verify the result.

Example 9

Factor $3x^2 - 7$.

Solution

$$3x^2 - 7 = (x\sqrt{3} - \sqrt{7})(x\sqrt{3} + \sqrt{7}).$$

∎∎∎

Example 10
Solve, by factoring, the quadratic equation $x^2 - 5 = 0$.

Solution
We have,

$$x^2 - 5 = 0$$
$$(x - \sqrt{5})(x + \sqrt{5}) = 0$$
$$x - \sqrt{5} = 0 \text{ or } x + \sqrt{5} = 0$$
$$x = \sqrt{5} \text{ or } x = -\sqrt{5}$$

Sometimes, we write $x = \pm\sqrt{5}$ for short.

∎∎∎

Very often, for reasons to be explained below, algebraic manipulations are performed on quotients involving radicals in the denominator. The object is to transform the given expression into an algebraically equivalent expression which does not contain any radical expressions in the denominator. This process is called *rationalizing the denominator*.

Rationalization

Consider the expression $\frac{3}{\sqrt{12}}$, we want to rationalize the denominator of this expression, that is, convert it to an equivalent expression not containing any radicals in its denominator. The procedure is very straightforward. Our objective is to transform the radicand of the denominator into a perfect square. There are many ways this may be accomplished, one such way is to transform the radicand of the denominator into the *smallest perfect square* which is a multiple of 12. Of course, that perfect square is 36. We proceed as follows.

$$\frac{3}{\sqrt{12}} = \frac{3}{\sqrt{12}}\frac{\sqrt{3}}{\sqrt{3}} = \frac{3\sqrt{3}}{\sqrt{36}} = \frac{3\sqrt{3}}{6} = \frac{\sqrt{3}}{2}$$

Note that to transform the denominator into the $\sqrt{36}$, we multiplied both the numerator and denominator by the $\sqrt{3}$. An alternate approach is to first simplify the denominator, and then find a multiple of 3 which is a perfect square, 9, and multiply numerator and denominator by the number needed to get this perfect square, illustrated as follows

$$\frac{3}{\sqrt{12}} = \frac{3}{\sqrt{4}\sqrt{3}} = \frac{3}{2\sqrt{3}} = \frac{3}{2\sqrt{3}}\frac{\sqrt{3}}{\sqrt{3}} = \frac{3\sqrt{3}}{2\cdot 3} = \frac{\sqrt{3}}{2}$$

Notice that the first approach is more direct. Alternately, both the numerator and denominator of the original expression could be multiplied by the denominator $\sqrt{12}$ and then simplify, this is the least desirable of the three methods, try it yourself to see why. To recapitulate, the most effective method is to find the smallest multiple of the radicand of the square root in the denominator and then modify the expression to obtain this perfect square.

Example 11

Rationalize the denominator and simplify $\dfrac{3}{\sqrt{8}}$.

Solution

The smallest multiple of 8 which is a perfect square is 16, and 16/8 = 2, so we proceed as follows

$$\frac{3}{\sqrt{8}} = \frac{3}{\sqrt{8}} \frac{\sqrt{2}}{\sqrt{2}} = \frac{3\sqrt{2}}{\sqrt{16}} = \frac{3\sqrt{2}}{4}$$

■■■

Example 12

Rationalize the denominator and simplify $\sqrt{\dfrac{12}{5}}$.

Solution

$$\sqrt{\frac{12}{5}} = \frac{\sqrt{12}}{\sqrt{5}} = \frac{\sqrt{12}}{\sqrt{5}} \frac{\sqrt{5}}{\sqrt{5}} = \frac{\sqrt{60}}{5} = \frac{\sqrt{4}\sqrt{15}}{5} = \frac{2\sqrt{15}}{5}$$

■■■

Example 13

Rationalize the denominator and simplify $\dfrac{\sqrt{10}\,\sqrt{2}}{\sqrt{6}\,\sqrt{3}}$.

Solution

$$\frac{\sqrt{10}\,\sqrt{2}}{\sqrt{6}\,\sqrt{3}} = \frac{\sqrt{20}}{\sqrt{18}} = \frac{\sqrt{20}}{\sqrt{18}}\frac{\sqrt{2}}{\sqrt{2}} = \frac{\sqrt{40}}{\sqrt{36}} =$$
$$\frac{\sqrt{4}\sqrt{10}}{6} = \frac{2\sqrt{10}}{6} = \frac{\sqrt{10}}{3}$$

The simplifications could have been done a little faster as follows:

$$\frac{\sqrt{10}\,\sqrt{2}}{\sqrt{6}\,\sqrt{3}} = \sqrt{\frac{20}{18}} = \sqrt{\frac{10}{9}} = \frac{\sqrt{10}}{\sqrt{9}} = \frac{\sqrt{10}}{3}$$

■■■

Example 14

Rationalize the denominator and simplify $\dfrac{4x^3}{\sqrt{12x}}$. (Assume $x > 0$.)

Solution

A multiple of 12 which is a perfect square is 36 and a multiply of x which is a perfect square is x^2.

Therefore, $36x^2$ is the radicand that we will obtain in the denominator.

$$\frac{4x^3}{\sqrt{12x}} = \frac{4x^3}{\sqrt{12x}}\frac{\sqrt{3x}}{\sqrt{3x}} = \frac{4x^3\sqrt{3x}}{\sqrt{36x^2}} = \frac{4x^3\sqrt{3x}}{6x} = \frac{2x^2\sqrt{3x}}{3}$$

■■■

The procedure used to rationalize the denominator of a monomial term involving a square roots may be easily extended to any monomial radical in a denominator. We illustrate with an example.

Example 15

Rationalize the denominator and simplify $\dfrac{4}{\sqrt[3]{4}}$.

Solution

We need to find the smallest multiple of 4 which is a perfect cube. Of course, that is 8, therefore we must transform the expression into an equivalent one having $\sqrt[3]{8}$ in the denominator.

$$\frac{4}{\sqrt[3]{4}} = \frac{4}{\sqrt[3]{4}}\frac{\sqrt[3]{2}}{\sqrt[3]{2}} = \frac{4\sqrt[3]{2}}{\sqrt[3]{8}} = \frac{4\sqrt[3]{2}}{2} = 2\sqrt[3]{2}$$

■■■

The examples considered above consisted of a monomial term in the denominator. Suppose now we have a binomial term in the denominator involving square roots, how do we rationalize such expressions? The key idea is the observation made above, that is, multiplication of a binomial expression involving square roots by its *conjugate* results in an expression without any square roots. We illustrate with examples.

Example 16

Rationalize the denominator and simplify $\dfrac{15}{6-\sqrt{3}}$.

Solution

The conjugate of the denominator is $6+\sqrt{3}$, therefore we multiply numerator and denominator by this conjugate and simplify.

$$\frac{15}{6-\sqrt{3}} = \frac{15}{(6-\sqrt{3})}\frac{(6+\sqrt{3})}{(6+\sqrt{3})} = \frac{15(6+\sqrt{3})}{36-3} = \frac{15(6+\sqrt{3})}{33} = \frac{5(6+\sqrt{3})}{11}$$

If desired, the 5 in the numerator may be distributed.
■■■

Example 17

Rationalize the denominator and simplify $\dfrac{\sqrt{3}+2\sqrt{5}}{\sqrt{3}-2\sqrt{5}}$.

Solution

Once again, we focus on the denominator and multiply both the numerator and denominator by the conjugate of the denominator.

$$\frac{\sqrt{3}+2\sqrt{5}}{\sqrt{3}-2\sqrt{5}} = \frac{(\sqrt{3}+2\sqrt{5})}{(\sqrt{3}-2\sqrt{5})}\frac{(\sqrt{3}+2\sqrt{5})}{(\sqrt{3}+2\sqrt{5})} = \frac{3+4\sqrt{15}+4\cdot5}{3-4\cdot5} =$$

$$\frac{23+4\sqrt{15}}{-17} = -\frac{23+4\sqrt{15}}{17}$$

■■■

Why is rationalization of a denominators so important? Before calculators were used, if we needed to approximate $\frac{4}{\sqrt{3}}$, it would require the division of 4 by 1.7320508. This is an easy division, but it is somewhat time consuming. On the other hand, consider $\frac{4}{\sqrt{3}} = 4\frac{\sqrt{3}}{3}$; to approximate this equivalent expression we need only multiply 4 by 1.7320508 and divide by 3. Multiplication by a decimal expression is usually faster than division by one. However, if we were using a calculator, it is just as easy to approximate the first expression as the second. Why then do we need to rationalize denominators? We really need it for expressions involving sums and differences of square roots in a denominator when we are performing an evaluation. Recall that zero divided by any non-zero number is 0, while a non-zero number divided by zero is undefined. What happens if we are evaluating an expression and we obtain zero divided by zero? Such an expression is called an *indeterminate* form and it may have meaning. In calculus, such expressions occur regularly and one tool used in evaluating such expressions is rationalization. Consider the expression

Indeterminate forms

$$\frac{h}{2-\sqrt{4+h}}$$

Try to evaluate it with your calculator when h is 0, you will get some kind of error message, depending on the calculator you use. When $h = 0$, both the numerator and denominator of this expression are zero. However, if we rationalize the denominator, we have

$$\frac{h}{2-\sqrt{4+h}} = \frac{h}{(2-\sqrt{4+h})}\frac{(2+\sqrt{4+h})}{(2+\sqrt{4+h})} =$$

$$\frac{h(2+\sqrt{4+h})}{4-(4+h)} = -\frac{h(2+\sqrt{4+h})}{h}$$

This last expression is algebraically equivalent to the original expression when h is not zero. Suppose, *before* we set h equal to zero we continue the simplification. Then we obtain

$$-\frac{h(2+\sqrt{4+h})}{h} = -(2+\sqrt{4+h})$$

If we *now* set $h = 0$, this last expression evaluates to - 4. It is precisely for calculations of the above type that we need to know how to rationalize denominators.

Exercise 18

Given the expression $\dfrac{h}{\sqrt{9+h}-3}$. (a) What happens if this expression is evaluated when $h=0$? (b) Rationalize the denominator of this expression and completely simplify the obtain an equivalent algebraic expression. (c) Evaluate the expression found in (b) when $h=0$.

Solution

(a) Setting $h=0$ results in an *indeterminate* form as both the numerator and denominator are 0.

(b)
$$\frac{h}{\sqrt{9+h}-3} = \frac{h}{(\sqrt{9+h}-3)} \frac{(\sqrt{9+h}+3)}{(\sqrt{9+h}+3)} =$$

$$\frac{h(\sqrt{9+h}+3)}{(9+h)-9} = \frac{h(\sqrt{9+h}+3)}{h} = \sqrt{9+h}+3$$

(c) evaluating $\sqrt{9+h}+3$ when $h=0$ yields 6.

■■■

Sometimes, in problems similar to the last one, we have to rationalize the *numerator* in order to perform the necessary evaluation, as the next example illustrates.

Example 19

Evaluate the expression $\dfrac{\sqrt{9+h}-3}{h}$ by first rationalizing the numerator and simplifying and then set $h=0$.

Solution

$$\frac{\sqrt{9+h}-3}{h} = \frac{(\sqrt{9+h}-3)}{h} \frac{(\sqrt{9+h}+3)}{(\sqrt{9+h}+3)} = \frac{9+h-9}{h(\sqrt{9+h}+3)} = \frac{h}{h(\sqrt{9+h}+3)} = \frac{1}{(\sqrt{9+h}+3)}$$

Setting $h=0$, the simplified expression evaluates to 1/6.

■■■

Exercise set 4

In Exercises 1 - 23 perform the indicated operation and simplify.

1. $3(4+\sqrt{2})$

2. $2(3-7\sqrt{5})$

3. $2(2\sqrt{3}-3\sqrt{5})$

4. $-5(4\sqrt{3}-2\sqrt{7})$

5. $2\sqrt{3}(2\sqrt{5}+3\sqrt{7})$

6. $\sqrt{2}(3\sqrt{2}-5\sqrt{3})$

7. $2\sqrt{3}(3\sqrt{6}-5\sqrt{3})$

8. $3\sqrt{5}(2\sqrt{5}-7\sqrt{10})$

9. $(3+\sqrt{3})(2+\sqrt{3})$

10. $(3-2\sqrt{3})(3+2\sqrt{5})$

11. $(2\sqrt{3}-4\sqrt{5})(3\sqrt{3}+2\sqrt{5})$

12. $(3\sqrt{7}-2\sqrt{3})(4\sqrt{7}+2\sqrt{5})$

13. $(3+2\sqrt{2})^2$

14. $(3-2\sqrt{2})^2$

15. $(2\sqrt{3}-3\sqrt{5})^2$

16. $(3\sqrt{5}+4\sqrt{6})^2$

17. $(3+2\sqrt{2})(3-2\sqrt{2})$

18. $(5+3\sqrt{6})(5-3\sqrt{6})$

19. $(3\sqrt{5}-2)(3\sqrt{5}+2)$

20. $(3\sqrt{5}-2\sqrt{2})(3\sqrt{5}+2\sqrt{2})$

21. $(2\sqrt{11}-3\sqrt{5})(2\sqrt{11}+3\sqrt{5})$

22. $(2-3\sqrt{2})^2(2+3\sqrt{2})^2$

23. $(3\sqrt{2}-4\sqrt{5})^2(3\sqrt{2}+4\sqrt{5})^2$

In exercises 24 -52, rationalize the denominator and simplify.

24. $\dfrac{3}{\sqrt{2}}$

25. $\dfrac{5}{\sqrt{10}}$

26. $\dfrac{9}{\sqrt{12}}$

27. $\dfrac{3}{\sqrt{6}}$

28. $\dfrac{14}{\sqrt{7}}$

29. $\dfrac{15}{\sqrt{18}}$

30. $\sqrt{\dfrac{5}{7}}$

31. $\sqrt{\dfrac{2}{3}}$

32. $\sqrt{\dfrac{5}{12}}$

33. $\sqrt{\dfrac{7}{18}}$

34. $\dfrac{\sqrt{20}}{\sqrt{5}}$

35. $\dfrac{\sqrt{18}}{\sqrt{3}}$

36. $\dfrac{\sqrt{15}}{\sqrt{5}}$

37. $\dfrac{4}{\sqrt{16x^3}}$ $(x>0)$

38. $\dfrac{5}{\sqrt{8x}}$ $(x>0)$

39. $\dfrac{6}{\sqrt[3]{2}}$

40. $\dfrac{12}{\sqrt[3]{9}}$

41. $\dfrac{24}{\sqrt[3]{16}}$

42. $\dfrac{8}{\sqrt[3]{16x^5}}$ $(x>0)$

43. $\dfrac{18}{\sqrt[4]{8}}$

44. $\dfrac{30}{\sqrt[4]{27}}$

45. $\dfrac{21}{\sqrt[4]{27x^2}}$ $(x > 0)$

46. $\dfrac{8}{3 - \sqrt{5}}$

47. $\dfrac{12}{2\sqrt{2} - 3}$

48. $\dfrac{3}{3\sqrt{2} - 2\sqrt{3}}$

49. $\dfrac{15}{4\sqrt{5} + 5\sqrt{3}}$

50. $\dfrac{39}{3\sqrt{7} + 2\sqrt{6}}$

51. $\dfrac{2 + 3\sqrt{5}}{2 - 3\sqrt{5}}$

52. $\dfrac{3 - 4\sqrt{3}}{3 + 4\sqrt{3}}$

53. Given the expression $\dfrac{h - 16}{\sqrt{h} - 4}$, simplify the expression by rationalization and after all simplification is completed, evaluate the expression for $h = 16$.

54. Given the expression $\dfrac{\sqrt{4 + h} - 2}{h}$, simplify the expression by rationalization and after all simplification is completed, evaluate the expression for $h = 0$.

55. Given the expression $\dfrac{2 - \sqrt{4 + h}}{2h}$, simplify the expression by rationalization and after all simplification is completed, evaluate the expression for $h = 0$.

56. Given the expression $\dfrac{h - 9}{\sqrt{h} - 3}$, simplify the expression by rationalization and after all simplification is completed, evaluate the expression for $h = 9$.

57. Given the expression $\dfrac{h}{2 - \sqrt{4 + h}}$, simplify the expression by rationalization and after all simplification is completed, evaluate the expression for $h = 0$.

Factor the given expression in exercises 58 - 63

58. $x^2 - 11$

59. $x^2 - 12$

60. $4x^2 - 15$

61. $12x^2 - 7$

62. $20x^2 - 9$

63. $18x^2 - 10$

In exercises 64 - 67, solve the quadratic equation for x by factoring.

64. $x^2 - 7 = 0$

65. $x^2 - 12 = 0$

66. $8x^2 - 3 = 0$

67. $27x^2 - 8 = 0$

POSTTEST 4- Time 15 minutes

Each question is worth 1 point.

Simplify each of the following expressions.

1. $3(5 + 2\sqrt{3}) - 3(-2 + 5\sqrt{3})$

2. $5\sqrt{2}(4 - 3\sqrt{10})$

3. $(7 - 2\sqrt{5})(6 + 3\sqrt{5})$

4. $(4\sqrt{6} - 3\sqrt{5})^2$

5. $(2\sqrt{7} - 5\sqrt{2})(2\sqrt{7} + 5\sqrt{2})$

6. Factor $x^2 - 20$

In each of the following, rationalize the denominator:

7. $\dfrac{16}{\sqrt{20}}$

8. $\dfrac{12\sqrt{10}\sqrt{30}}{\sqrt{2}\sqrt{6}}$

9. $\dfrac{18}{\sqrt[3]{9}}$

10. $\dfrac{34}{3\sqrt{5} - 2\sqrt{7}}$

44

(Notes)

5. Complex Numbers

» **Imaginary Number i**
» **Complex Numbers**

PRETEST 5 Time 10 minutes

All answers should be given in simplest form. Each question is worth one point.

Simplify:

1. $\sqrt{-25}\sqrt{-9}$

2. $\sqrt{-24}$

3. $(5 + 7i) - (2 - 3i)$

4. $(3 + 5i)(2 - 3i)$

5. $(3 - 2i)^2$

6. $(5 + 7i)(5 - 7i)$

7. $\dfrac{5}{3i}$

8. $\dfrac{58}{5 - 2i}$

Solve for x:

9. $x^2 + 4 = 0$

10. $18x^2 + 25 = 0$

Consider the quadratic equation

$$x^2 + 1 = 0 \tag{1}$$

It should be clear that this equation has no *real* solution. Why? Suppose it had one. Certainly x could not be zero as zero does not check the equation. Furthermore, x cannot be a positive or negative number because squaring any such number and adding 1 results in a number larger than one. Therefore, no *real* number x can solve this equation.

Imaginary Number i

Let us define a new number (which is not a real number) that is the solution to this equation and denote this number by i. Thus, if this number is indeed the solution to (1), then we must have

$$i^2 + 1 = 0$$

or equivalently,

$$i^2 = -1$$

or

45

$$i = \sqrt{-1}$$

This number i, when it was first defined, was called an *imaginary* number.

We remark that the quadratic equation given by (1) has in fact two solutions, namely $x = i$ and $x = -i$. (Squaring either of these solutions yields -1, and $-1 + 1 = 0$, as required.)

Observe that we can now allow the radicand of a square root to be negative. For example, we have that $\sqrt{-4} = \sqrt{4}\sqrt{-1} = 2i$, and $\sqrt{-8} = \sqrt{-1}\sqrt{8} = i\sqrt{4}\sqrt{2} = 2i\sqrt{2} = 2\sqrt{2}\,i$. However, we must be careful when simplifying the product of two square roots when each radicand is negative. There is a subtle but important observation to be made. Recall that if a is positive, $\sqrt{a}\sqrt{a} = a$. This result is true when $a = -1$, as $\sqrt{-1}\sqrt{-1} = i\cdot i = i^2 = -1$. As a consequence of this, we have

$$\sqrt{-9}\sqrt{-4} = \sqrt{-1}\sqrt{9}\sqrt{-1}\sqrt{4} = 3i\cdot 2i = 6i^2 = -6.$$

But observe that the following simplification is incorrect,

$$\sqrt{-9}\sqrt{-4} \neq \sqrt{(-9)(-4)} = \sqrt{36} = 6$$

Thus, the square root of a product is not equal to the product of the square roots when both radicands are negative.

Example 1
Simplify each of the following (a) $\sqrt{-18}$ (b) $\sqrt{-12}\sqrt{-5}$.

Solution
(a) $\sqrt{-18} = \sqrt{18}\sqrt{-1} = \sqrt{9}\sqrt{2}\,i = 3\sqrt{2}\,i.$

(b) $\sqrt{-12}\sqrt{-5} = \sqrt{-1}\sqrt{12}\sqrt{-1}\sqrt{5} = i\sqrt{12}\,i\sqrt{5} = i^2\sqrt{60} = -\sqrt{4}\sqrt{15} = -2\sqrt{15}.$
■■■

Complex numbers

By a *complex* number we mean any number of the form $a + bi$ where a and b are real numbers. If a is zero then the complex number is imaginary, and if b is zero then the complex number is real. $3 + 2i$, $6 - 5i$, $5 + 2\sqrt{3}\,i$ are each examples of a complex number. We can add, subtract, multiply and divide complex numbers in the same way that we do real numbers.

Example 2
Simplify each of the following (a) $(4+2i)-(3-5i)$ (b) $(2+3i)(-5+7i).$

Solution
(a) $(4+2i)-(3-5i) = 4+2i-3+5i = 1+7i.$

(b) $(2+3i)(-5+7i) = -10+14i-15i+21i^2 = -10-i+21(-1) = -31-i.$
■■■

Since a complex number has the same form as an irrational number we can also *rationalize* denominators containing complex numbers, using the same techniques as with irrational numbers. Consider the complex numbers $a + bi$ and $a - bi$ they are *complex conjugates* of each other, multiplying them together results in a real number, that is

$$(a + bi)(a - bi) = a^2 - b^2i^2 = a^2 - b^2(-1) = a^2 + b^2$$

(Note that in the multiplication, the inner and outer terms canceled.)

Example 3

Rationalize the denominator and simplify the given expression, (a) $\dfrac{7}{2i}$ (b) $\dfrac{100}{3+4i}$.

Solution

(a) $\dfrac{7}{2i} = \dfrac{7}{2i} \cdot \dfrac{i}{i} = \dfrac{7i}{2i^2} = \dfrac{7i}{-2} = -\dfrac{7i}{2}$.

(b) $\dfrac{100}{3+4i} = \dfrac{100}{(3+4i)} \cdot \dfrac{(3-4i)}{(3-4i)} = \dfrac{100(3-4i)}{9-16i^2} = \dfrac{100(3-4i)}{9+16} = \dfrac{100(3-4i)}{25} = 4(3-4i) = 12 - 16i$.

■■■

Just as we can use the notion of real conjugates to factor binomials of the form $ax^2 - b$, we can use complex conjugates to factor binomials of the $ax^2 + b$, in fact, we have, if a and b are real numbers,

$$ax^2 + b = (x\sqrt{a} - \sqrt{b}\,i)(x\sqrt{a} + \sqrt{b}\,i)$$

We ask you to verify this result by multiplying out of the right hand side of the equation. We illustrate this identity in the next examples.

Example 4
Factor $x^2 + 4$.

Solution

$$x^2 + 4 = (x + 2i)(x - 2i)$$

■■■

Example 5
Factor $x^2 + 8$.

Solution

$$x^2 + 8 = (x + \sqrt{8}\,i)(x - \sqrt{8}\,i) = (x + 2\sqrt{2}\,i)(x - 2\sqrt{2}\,i)$$

(Remember, $\sqrt{8} = \sqrt{4}\sqrt{2} = 2\sqrt{2}$.)

■■■

Example 6
Factor $12x^2 + 5$.

Solution

$$12x^2 + 5 = (x\sqrt{12} + \sqrt{5}\,i)(x\sqrt{12} - \sqrt{5}\,i) = (2x\sqrt{3} + i\sqrt{5})(2x\sqrt{3} - i\sqrt{5})$$

(Remember, $\sqrt{12} = \sqrt{4}\sqrt{3} = 2\sqrt{3}$.)

■■■

Example 7
Solve the quadratic equation $12x^2 + 5 = 0$.

Solution
Following the solution from the previous example, we have

$$(2x\sqrt{3} + i\sqrt{5})(2x\sqrt{3} - i\sqrt{5}) = 0,$$
$$2x\sqrt{3} + i\sqrt{5} = 0 \text{ or } 2x\sqrt{3} - i\sqrt{5} = 0$$
$$2x\sqrt{3} = -i\sqrt{5} \text{ or } 2x\sqrt{3} = i\sqrt{5}$$
$$x = -i\frac{\sqrt{5}}{2\sqrt{3}} \text{ or } x = i\frac{\sqrt{5}}{2\sqrt{3}}$$
$$x = -i\frac{\sqrt{15}}{6} \text{ or } x = -i\frac{\sqrt{15}}{6}$$
$$x = \pm i\frac{\sqrt{15}}{6}$$

(Remember, $\dfrac{\sqrt{5}}{2\sqrt{3}} = \dfrac{\sqrt{5}}{2\sqrt{3}}\dfrac{\sqrt{3}}{\sqrt{3}} = \dfrac{\sqrt{15}}{2\cdot 3} = \dfrac{\sqrt{15}}{6}$.)

■■■

Exercise set 5
In Exercise 1 - 10 rewrite the expression without any negative signs in the radical.

1. $\sqrt{-25}$

2. $\sqrt{-36}$

3. $\sqrt{-18}$

4. $\sqrt{-27}$

5. $\sqrt{-9}\sqrt{-16}$

6. $\sqrt{-12}\sqrt{-3}$

7. $3\sqrt{-8} + 2\sqrt{-2}$

8. $\dfrac{\sqrt{-25}}{\sqrt{-36}}$

9. $\sqrt{-2}\sqrt{-3}\sqrt{-6}$

10. $\sqrt{-2}\sqrt{-3}\sqrt{-8}$

In exercises 11 -34, write the answer in the form $a + bi$.

11. $(3 + 2i) + (5 - 7i)$

12. $(2 - 3i) + (8 + 9i)$

13. $3(2 + 3i) - 4(5 - 6i)$

14. $-4(6 - 7i) + 3(3 + 4i)$

15. $(3 + 2i)(5 - 3i)$

16. $(7 - 3i)(2 + 5i)$

17. $(2 + 3i)^2$

18. $(5 - 3i)^2$

19. $(2 + 5i)(2 - 5i)$

20. $(3 - 7i)(3 + 7i)$

21. $(3\sqrt{2} - 2i)(3\sqrt{2} + 2i)$

22. $(2\sqrt{2} - 3\sqrt{5}i)(2\sqrt{2} + 3\sqrt{5}i)$

23. $\dfrac{15}{2i}$

24. $\dfrac{9}{5i}$

25. $\dfrac{3 - 2i}{4i}$

26. $\dfrac{9 + 7i}{3i}$

27. $\dfrac{15}{3\sqrt{12}i}$

28. $\dfrac{20}{3\sqrt{5}i}$

29. $\dfrac{26}{3 - 2i}$

30. $\dfrac{-58}{2 + 5i}$

31. $\dfrac{42}{2\sqrt{3} - 3i}$

32. $\dfrac{3 + 2i}{3 - 2i}$

33. $\dfrac{5 + 7i}{4 - 3i}$

34. $\dfrac{8 + 5i}{6 + 2i}$

In exercises 35 - 40, factor the given binomial.

35. $x^2 + 4$

36. $x^2 + 9$

37. $x^2 + 8$

38. $4x^2 + 25$

39. $9x^2 + 16$

40. $12x^2 + 25$

In exercises 41 - 45, solve the given quadratic equation for x using factoring.

41. $x^2 + 25 = 0$

42. $16x^2 + 9 = 0$

43. $5x^2 + 8 = 0$

44. $12x^2 + 25 = 0$

45. Show that i^N where N is any integer, positive or negative, will assume one of the following values: -1, 1, -i, i.

In exercises 46 - 53, simplify the expression using exercise 45.

46. i^{23}

47. i^{28}

48. i^{406}

50

49. i^{403}

50. $\left(\sqrt{2}\, i\right)^{10}$

51. $\left(\sqrt{3}\, i^3\right)^4$

52. $\dfrac{12}{7i^{47}}$

53. $4i^{-39}$

POSTTEST5 Time 10 minutes

All answers should be given in simplest form.   Each question is worth one point.

Simplify:

1. $\sqrt{-49}\sqrt{-16}$ 2. $\sqrt{-45}$ 3. $(7 - 8i) - (6 - 9i)$

4. $(4 +7i)(3 - 5i)$ 5. $(4 - 3i)^2$ 6. $(6 - 2i)(6 + 2i)$

7. $\dfrac{12}{5i}$ 8. $\dfrac{75}{4+3i}$

Solve for x:

9. $x^2 + 9 = 0$ 10. $45x^2 + 8 = 0$

6. Fractional Exponents

» Fractional Roots
» Equations with Fractional Exponents
» Products of Radicals
» Compound Interest
» Rule of 72
» Simple Pendulum

PRETEST 6- Time 10 minutes

Each question is worth one point.

Simplify

1. $27^{\frac{1}{3}}$ 　　　　 2. $8^{-\frac{2}{3}}$ 　　　　 3. $27^{\frac{4}{3}}$ 　　　 4. $16^{-\frac{3}{4}}$

Solve for x

5. $x^{\frac{2}{3}} = 9$ 　　　 6. $3x^{\frac{3}{4}} = 24$ 　　　 7. $\sqrt[4]{5x+6} = 2$

8. Write as a single radical, $\sqrt{2}\,\sqrt[3]{2}$

9. If \$1,000 is invested in an account yielding 8% interest compounded quarterly, to what will it accumulate at the end of 5 years?

10. What interest rate compounded annually will double an investment in 9 years?

Fractional Roots

In this section we shall first define what is meant by a fractional exponent and use it to solve some elementary radical equations. In Section 13, we shall use the methods of this section to solve more complicated radical equations involving square roots.

We want to define expressions like $4^{1/3}$, $8^{1/6}$, $19^{2/3}$ and so on. We have the following definition.

Let $b > 0$ and n be a positive integer, then we define

$$b^{\frac{1}{n}} = \sqrt[n]{b} \tag{1}$$

First, we should demonstrate that this definition is consistent with our rules for exponents. Observe that

$$4^{\frac{1}{3}}4^{\frac{1}{3}}4^{\frac{1}{3}} = 4^1 = 4$$

this is as it should be, since $4^{\frac{1}{3}} = \sqrt[3]{4}$ and

$$\sqrt[3]{4}\sqrt[3]{4}\sqrt[3]{4} = 4$$

Example 1

Simplify each of the following (a) $8^{\frac{1}{3}}$ (b) $16^{\frac{1}{4}}$ (c) $27^{-\frac{1}{3}}$

Solution

(a) $8^{\frac{1}{3}} = \sqrt[3]{8} = 2$

(b) $16^{\frac{1}{4}} = \sqrt[4]{16} = 2$

(c) $27^{-\frac{1}{3}} = \dfrac{1}{27^{\frac{1}{3}}} = \dfrac{1}{\sqrt[3]{27}} = \dfrac{1}{3}$

◼◼◼

Observe in (c) of the previous exercise, we used the law regarding negative exponents, namely

$$b^{-n} = \frac{1}{b^n}$$

With our definition of $b^{\frac{1}{n}}$ we can go one step further. What do we mean by $b^{\frac{m}{n}}$? We will want our definition to be consistent with the laws of exponents that you have already learned. Using the law

$$(b^m)^n = b^{mn}$$

let us see what the definition should be. We assume as above, that $b > 0$ and m and n are integers and $n \neq 0$. We want this rule to hold true for all exponents, so in particular,

$$b^{\frac{m}{n}} = \left(b^{\frac{1}{n}}\right)^m = \left(\sqrt[n]{b}\right)^m \qquad (2)$$

We illustrate this rule in the next example.

Example 2

Evaluate (a) $16^{\frac{3}{4}}$ (b) $27^{\frac{2}{3}}$ (c) $8^{-\frac{2}{3}}$

Solution

(a) $16^{\frac{3}{4}} = \left(\sqrt[4]{16}\right)^3 = 2^3 = 8$

(b) $27^{\frac{2}{3}} = \left(\sqrt[3]{27}\right)^2 = 3^2 = 9$

(c) $8^{-\frac{2}{3}} = \dfrac{1}{8^{\frac{2}{3}}} = \dfrac{1}{\left(\sqrt[3]{8}\right)^2} = \dfrac{1}{2^2} = \dfrac{1}{4}$

■ ■ ■

Equations involving fractional exponents can be solved by isolating the term containing the power with the variable in its base, and then raising both sides of the equation to the appropriate power. For example, to solve the equation

$$x^{\frac{3}{2}} = 27$$

Equations with Fractional Exponents

We raise both sides to the power 2/3, which is the reciprocal of the power 3/2. The we solve the resulting equation.

$$\left(x^{\frac{3}{2}}\right)^{\frac{2}{3}} = 27^{\frac{2}{3}}$$
$$x^1 = 27^{\frac{2}{3}}$$
$$x = \left(\sqrt[3]{27}\right)^2 = 3^2 = 9$$

Observe that by raising both sides of the equation to the reciprocal of the power of the term involving the unknown results in the term containing the unknown having power one.

Example 3

Solve the equation $\sqrt[3]{4x+3} = 3$.

Solution
First, we rewrite the equation is exponential form,

$$\sqrt[3]{4x+3} = (4x+3)^{\frac{1}{3}} = 3.$$

Since the reciprocal of 1/3 is 3, we raise both sides to the third power and then solve for x.

$$\left((4x+3)^{\frac{1}{3}}\right)^3 = 3^3$$
$$4x+3 = 27$$
$$4x = 24$$
$$x = 6$$

this value for x checks the original equation therefore it is the solution.

■ ■ ■

Observe that we just solved a radical equation involving a cube root. We really did not need the notion of a fractional exponent to do so, we could have just cubed both sides of the equation. However, thinking of any radical as a fractional power just gives a more uniform approach to solving equations with them, namely, isolate the fractional term, raise each side of the equation to the reciprocal power of the fractional exponent and solve.

One remark, if we were totally consistent, we would write $\sqrt[2]{}$ instead of $\sqrt{}$. However, when dealing with square roots, the index 2 is understood and is not written.

Products of Radicals

Suppose we have products of radicals with different indices, each containing the same radicand. We can now use our knowledge of fractional exponents to simplify such expressions. The next example illustrates how this may be accomplished.

Exercise 4

Write $\sqrt{2} \cdot \sqrt[3]{2}$ as a single radical.

Solution
We rewrite the given expression using fractional exponents, use the ordinary rules for multiplying exponents with the same base, and then convert the expression back to a radical.

$$\sqrt{2} \cdot \sqrt[3]{2} = 2^{\frac{1}{2}} 2^{\frac{1}{3}} = 2^{\frac{1}{2}+\frac{1}{3}} = 2^{\frac{5}{6}} = \sqrt[6]{2^5} = \sqrt[6]{32}$$

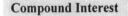

Compound Interest

In applications, examples involving fractional exponents arise very frequently. You may recall the formula for compound interest. If a principal P is deposited into an account giving interest at the rate r compounded n times per year, then the accumulation A at the end of t years is given by the formula

$$A = P(1 + \frac{r}{n})^{nt} \qquad \text{(3)}$$

We illustrate how we may need fractional exponents in finding the interest rate.

Example 5
$1,500 is invested into an account paying interest compounded monthly for 5 years at which time it accumulates to $1,675. How much interest, compounded monthly, did the account earn?

Solution
We are given $A = 1675$, $P = 1500$, $t = 5$ and $n = 12$, we must find r. Using (3), we have

$$1675 = 1500(1 + r/12)^{60}$$

We need to isolate the exponential term so we divide both sides by 1500, yielding

$$1.116666 = (1 + r/12)^{60}$$

we raise each side to the 1/60 power and then solve for r. A calculator will be needed to help with the calculations.

$$1.116666^{\frac{1}{60}} = \left((1 + \frac{r}{12})^{60} \right)^{\frac{1}{60}}$$

Using a calculator, we find that

$$1.116666^{\frac{1}{60}} = 1.0018408$$

therefore

$$1.0018408 = 1 + \frac{r}{12}$$

$$\frac{r}{12} = .0018408$$

$$r = 0.02208992$$

or to the nearest one-hundredth of a percent, $r = 2.21\%$.

■■■

We remark that different calculators do powers differently. On most graphical calculators, entering the following would perform the required 60th root; 1.116666^(1/60) = . Sometimes, enter is pressed in place of the equal sign. Non-graphical calculators have an exponential key given usually as either as X^y or Y^x. on these calculators, you would press the following keys: 1.116666 X^y (1/60) = .

Example 6
What interest rate, compounded quarterly, will result in an investment doubling in 6 years?

Solution
We may choose any amount as our principal, say $1. For it to double means that the accumulation will be $2. Substitution into (1) yields $2 = 1(1 + r/4)^{24}$, where r is the desired rate. To solve the equation $2 = (1 + r/4)^{24}$, we take the 24th root of each side of the equation. Thus,

$$2^{1/24} = \left[(1 + \frac{r}{4})^{24}\right]^{1/24}$$

or

$$2^{1/24} = 1 + r/4.$$

We use our calculator to find that $2^{1/24} = 1.029302237$. Thus,

$$1.029302237 = 1 + r/4,$$

or

$$r/4 = 0.029302237$$

or

$$r = 0.117208946.$$

To the nearest one-hundredth of a percent, we have $r = 11.72\%$ as the rate at which the investment will double in 6 years.

■■■

Rule of 72

There is a rule, well known to the financial community, which yields an approximation to finding the "doubling time" for an investment earning compound interest. It is called *The Rule of 72*, which states that the time, in years, for an investment to double itself earning compound interest at the rate $r\%$ is given by

$$t \approx \frac{72}{r}$$

Example 7
An investment earns interest, compounded annually at a rate of 12%. Determine its approximate doubling time.

Solution
Using the rule of 72, we have

$$t \approx \frac{72}{12} = 6 \text{ years}$$

■■■

You might be wondering how to find the time exactly if we are given the principal and interest rate. A knowledge of logarithms is needed to find the time exactly.

Simple Pendulum	As another example of how fractional exponents arise, consider the problem of finding the period T, of a simple pendulum of length l. In physics, it can be shown that the formula relating the period and length is

$$T = 2\pi \sqrt{\frac{l}{g}} \qquad (4)$$

where π is the famous constant equal to the ratio of the circumference of any circle to its diameter, and to five decimal places is 3.14159, g is the acceleration due to gravity and is approximately equal to 32 feet per second per second (or in metric units, 9.8 meters per second per second). Calculations in such problems will often require a calculator.

Example 8
How long (to the nearest inch) does a pendulum have to be if its period is 15 seconds?

Solution
Observe from the units, that if g is given in feet per second per second, T in seconds, then l will be given in feet. We have, after substituting into (4)

$$15 = 2\pi \sqrt{\frac{l}{32}}$$

$$2.3873241 = \left(\frac{l}{32}\right)^{\frac{1}{2}}$$

$$2.3873241^2 = \left(\left(\frac{l}{32}\right)^{\frac{1}{2}}\right)^2$$

$$5.69931658 = \frac{l}{32}$$

$$l = 182.3781305 \text{ feet}$$

0.3781305 feet = 0.3781305(12 inches) = 4.53 inches. Thus, the pendulum should be 182 feet 5 inches long.

■■■

Exercise set 6
In exercises 1 - 34, simplify the expression. There should be no negative exponents in any answer.

1. $32^{\frac{1}{5}}$

2. $32^{-\frac{1}{5}}$

3. $-32^{\frac{1}{5}}$

4. $-32^{-\frac{1}{5}}$

5. $27^{\frac{1}{3}}$

6. $-27^{\frac{1}{3}}$

7. $(-27)^{\frac{1}{3}}$

8. $-(-27)^{\frac{1}{3}}$

9. $(-27)^{-\frac{1}{3}}$

10. $16^{\frac{3}{4}}$

11. $16^{-\frac{3}{4}}$

12. $27^{\frac{5}{3}}$

13. $27^{-\frac{5}{3}}$

14. $(-125)^{-\frac{1}{3}}$

15. $8^{-\frac{2}{3}}$

16. $-8^{\frac{2}{3}}$

17. $9^{\frac{5}{2}}$

18. $16^{\frac{5}{4}}$

19. $(-8)^{-\frac{4}{3}}$

20. $(-32)^{\frac{3}{5}}$

21. $\dfrac{16^{-\frac{3}{4}}}{27^{-\frac{2}{3}}}$

22. $16^{\frac{5}{4}} 9^{-\frac{3}{2}}$

23. $b^{\frac{3}{4}} b^{\frac{5}{8}}$

24. $b^{\frac{3}{4}} b^{\frac{2}{3}}$

25. $\dfrac{b^{\frac{5}{4}}}{b^{\frac{1}{2}}}$

26. $(a^3)^{\frac{4}{3}}$

27. $\left(a^{\frac{1}{4}}\right)^8$

28. $\left(x^6 y^9\right)^{\frac{1}{3}}$

29. $\left(4x^6 y^8\right)^{\frac{1}{2}}$

30. $\left(-27x^9 y^3\right)^{\frac{2}{3}}$

31. $\left(\dfrac{x^8}{y^2}\right)^{\frac{1}{2}}$

32. $\left(\dfrac{x^8}{y^2}\right)^{-\frac{1}{2}}$

33. $\left(\dfrac{4x^4}{9y^6}\right)^{\frac{3}{2}}$

34. $\left(\dfrac{8x^6}{27y^{12}}\right)^{\frac{2}{3}}$

In exercises 35 - 47 solve for the unknown.

35. $x^{\frac{1}{3}} = 3$

36. $x^{\frac{1}{4}} = 2$

37. $5x^{\frac{2}{3}} = 125$

38. $\sqrt[3]{5x+3} = 2$

39. $\sqrt[4]{9x-18} = 3$

40. $3(1+3x)^{\frac{1}{5}} = 6$

41. $\left(1 - \dfrac{x}{2}\right)^4 = 16$

42. $\left(1 + \dfrac{x}{4}\right)^5 = 32$

43. $2\left(3 - \dfrac{2x}{3}\right)^4 = 162$

44. $5\left(7 + \dfrac{3}{4}x\right)^{\frac{2}{3}} + 10 = 55$

45. $6\left(4 + \dfrac{5}{3}x\right)^{\frac{3}{4}} - 2 = 46$

46. $\dfrac{1}{3}\left(\dfrac{5}{6}x - 3\right)^{\frac{5}{4}} - 2 = 79$

47. $\dfrac{1}{3}(7 - 2x)^{\frac{2}{3}} - 2 = 10$

In exercises 48- 56, write the given expression as a single radical.

48. $\sqrt[4]{5}\sqrt{5}$

49. $\sqrt[6]{3}\sqrt[3]{3}$

50. $\sqrt[8]{4}\sqrt[4]{4}$

51. $\sqrt[5]{2}\sqrt[3]{2}$

52. $\dfrac{\sqrt[4]{2}}{\sqrt[6]{2}}$

53. $\dfrac{\sqrt[3]{3}}{\sqrt[4]{3}}$

54. $\dfrac{\sqrt[4]{5}}{\sqrt[5]{5}}$

55. $\sqrt{2}\sqrt[3]{2}\sqrt[4]{2}$

56. $b^{\frac{m}{n}} b^{\frac{r}{s}}$

57. If \$500 accumulates to \$750 in 5 years, at what rate is the interest compounded monthly?

58. If \$275 accumulates to \$325 in 4 years, at what rate is the interest compounded quarterly?

59. If money doubles itself in ten years, at what rate is the interest compounded semi-annually?

60. If money triples itself in 30 years, at what rate is the interest compounded quarterly?

61. $1250 is deposited into an account for 5 years and accumulates to $1324. Assuming the interest is compounded annually, find the interest rate.

62. $250 is deposited into an accounted yielding interest compounded semi-annually. If at the end of 13 years it accumulates to $525, what is the semi-annual interest rate?

63. $6000 is invested for 7 years at 5.8% compounded quarterly. To what amount will it accumulate?

64. $86,746 is invested for 6½ years at 6.46% compounded semiannually. To what amount will it accumulate?

65. An investment accumulated to $525 in 3 years while earning interest at 5.25% compounded monthly. What was the original investment?

66. If you want to earn a rate of 7% annually on your investments, what would you pay today for a U.S. Government note that will pay $25,000, eight years from today?

67. Mr. Lorentz claims that his stock investments yield a rate of 12.2% compounded semiannually. If his money was invested nine and one-half years ago, and his investments are now worth $145,000, how much did he invest originally?

68. If $1200 invested six years ago, has accumulated to $1900, what rate compounded annually did it earn?

69. If $2500 invested ten years ago is today worth $4900, what rate compounded monthly did it earn?

70. If $2200 invested nine years ago, has accumulated to $3900, what rate compounded semiannually did it earn?

71. If $36,000 invested ten years ago is today worth $78,000, what rate compounded quarterly did it earn?

72. Verify the accuracy of the Rule of 72 by finding the rates at which money will double in 7 years, 8 years, and 10 years.

73. $7245 is invested for 8 years. Determine the accumulation if it earns interest at a rate of 8% compounded: (a) annually; (b) semiannually; (c) quarterly; (d) monthly; (e) daily; (f) hourly.

74. $9345 is invested for 6 years. Determine the accumulation if it earns interest at a rate of 6% compounded: (a) annually; (b) semiannually; (c) quarterly; (d) monthly; (e) daily; (f) hourly.

75. Approximately, how long does it take $1500.58 to double if it earns 6% interest, compounded monthly?

76. HDS Investment Company offers 8% compounded quarterly on investments. WBG Equity Trust advertises that it will match all competitors. However, for bookkeeping purposes, it will give interest monthly. What rate must it offer monthly to match HDS's 8% offering?

77. What interest rate, r, compounded m times per year is equivalent to an interest rate, s, compounded j times per year?

78. John will invest $100 today and $200 six months from today. In 2 years he will withdraw all his money. If the money was earning interest at a rate of 6.27% compounded: (a) monthly; (b) daily; how much will he withdraw?

79. How long must a pendulum be, to the nearest inch, if its period is 30 seconds? (b) 60 seconds?

80. How long must a pendulum be, to the nearest inch, if its period is 20 seconds? (b) 40 seconds?

81. What effect does quadrupling the length of a pendulum have on its period?

82. If the acceleration due to gravity on the Moon is about 1/5 that of the Earth's how long must a pendulum be on the Moon if its period is to be 15 seconds?

83. The cube root of the sum of some number and 7 is 2, find the number.

84. The cube root of 13 less than four times some number is 3, find the number.

85. The cube root of 12 diminished by 5 times some number is 4, find the number.

86. The sum of 5 times a number and 12 raised to the fourth power is 16, find the number.

87. Show that $\left(a^M\right)^{\frac{1}{N}}$ is an alternate definition of $a^{\frac{M}{n}}$.

88. Using the alternate definition given in the previous exercise, explain, by using it in computing the following, why it is not always the best way to simplify. (a) $16^{\frac{3}{4}}$ (b) $64^{\frac{5}{2}}$.

89. Show that the compound interest formula may be rewritten as $P = A(1 + \frac{r}{n})^{-nt}$. Why would this form be useful?

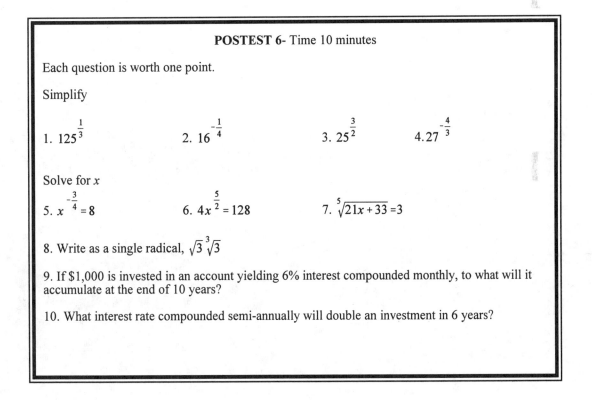

POSTEST 6- Time 10 minutes

Each question is worth one point.

Simplify

1. $125^{\frac{1}{3}}$　　　　2. $16^{-\frac{1}{4}}$　　　　3. $25^{\frac{3}{2}}$　　　　4. $27^{-\frac{4}{3}}$

Solve for x

5. $x^{-\frac{3}{4}} = 8$　　　　6. $4x^{\frac{5}{2}} = 128$　　　　7. $\sqrt[5]{21x + 33} = 3$

8. Write as a single radical, $\sqrt{3}\sqrt[3]{3}$

9. If $1,000 is invested in an account yielding 6% interest compounded monthly, to what will it accumulate at the end of 10 years?

10. What interest rate compounded semi-annually will double an investment in 6 years?

60

(Notes)

7. A REVIEW OF FACTORING

> » *GCF*
> » **Monomial Factors**
> » **Factoring Trinomials**
> » **Prefect Squares**
> » **Conjugates - The Difference of Two Squares**
> » **Factoring by Grouping**
> » **Quadratic Equations and Factoring**

PRETEST 7- Time 15 minutes

Each question is worth one point. Completely Factor each expression

1. $16x^2y^3 + 24x^2y^4$

2. $12x^{\frac{3}{2}} + 18x^{\frac{1}{2}}$

3. $9x^3(x^2 - 1)^{15} - 12x^5(x^2 - 1)^{14}$

4. $2x\sqrt{x^2 - 1} + \dfrac{x^3}{\sqrt{x^2 - 1}}$

5. $12x^2 - 31x + 20$

6. $20x^2y^2 - 9xy - 20$

7. $40x - 20x^2 - 15$

8. $4x^2 - 25$

9. $9x^2y^2 - 49z^2$

10. $3ax + 12a - 2x - 8$

Factoring is essentially a reversal of the distributive property which states

$$a(b + c) = ab + ac$$

GCF

Basically, to factor an expression means to rewrite it, using the distributive property, with the *largest* terms extracted, often called the greatest common factor (*GCF*). We illustrate the process through examples.

Example 1
Factor $12x + 15y$.

Solution

The *GCF* for 12 and 15 is 3, (3 goes into 12 4 times and into 15 5 times, so we have)

$$12x + 15y = 3(4x + 5y)$$

■■■

Note that we can always check our answer by distribution of the 3; we must, of course, obtain the original expression.

In Example 1, it was straight-forward to determine the *GCF*. When integers are larger, one can use trial an error, prime factorization, or the following variation: consider finding the *GCF* of integers, say 12 and 15. We write

$$12 \qquad 15 \qquad 3$$

Find any common factor of 12 and 15, and put it in the right column, divide this factor into both 12 and 15, putting their quotients in the row below them.

$$
\begin{array}{ccc}
12 & 15 & 3 \\
4 & 5 &
\end{array}
$$

We repeat the process; find any common factor of 4 and 5, since they are relatively prime, that is, their only positive factor is 1, we are done. Put the 1 in the right column. Now multiply the numbers in the right column together to obtain the *GCF*.

$$
\begin{array}{ccc}
12 & 15 & 3 \\
4 & 5 & 1
\end{array}
$$

To summarize: continue the process until the terms in the last row are relatively prime, that is, their only positive factor is 1; when that occurs, the product of the common factors in the last column is the *GCF*.

As another illustration, we find the *GCF* of 24 and 32.

$$24 \qquad 32$$

We choose any common factor of 24 and 32, say 4, which goes into these number 6 and 8 times respectively, then we have

$$
\begin{array}{ccc}
24 & 32 & 4 \\
6 & 8 &
\end{array}
$$

Next, we find a common factor of 6 and 8, 2, which goes into these number 3 and 4 times respectively, so we have

$$
\begin{array}{ccc}
24 & 32 & 4 \\
6 & 8 & 2 \\
3 & 4 &
\end{array}
$$

We repeat the process, but now since 3 and 4 are relatively prime, that is, only 1 goes into them, we are done.

$$
\begin{array}{ccc}
24 & 32 & 4 \\
6 & 8 & 2 \\
3 & 4 & 1
\end{array}
$$

The product of the numbers in the last column 4, 2, and 1 is 8 which is the *GCF*. Note, that the larger the factor you begin with, the fewer the steps. Repeat this example using 8 as your first factor.

Example 2
Factor $180x^2 + 72x^3$

Solution
We first observe that the *GCF* of 180 and 72 is 36, thus,

$$180x^2 + 72x^3 = 36(5x^2 + 2x^3)$$

Now the greatest common factor of x^2 and x^3 is x^2, therefore, we have

$$180x^2 + 72x^3 = 36x^2(5 + 2x)$$

■■■

Observe that the *GCF* of x^m and x^n where $m < n$ is x^m. Thus, the *GCF* of x^5 and x^8 is x^5, the *GCF* of x^{-2} and x^1 is x^{-2}, and the *GCF* of $x^{1/2}$ and $x^{3/4}$ is $x^{1/2}$. We could, of course, use the procedure used above with integers, but with exponents, this observation will most quickly yield the *GCF*.

Example 3
Factor $15x^5y^7z^4 + 20x^7y^3z^4$

Solution
We have

$$15x^5y^7z^4 + 20x^7y^3z^4 = 5x^5y^3z^4(3y^4 + 4x^2)$$

■■■

Example 4
Factor $12x^{-1/2} + 8x^{1/2}$

Solution
The *GCF* of $x^{-1/2}$ and $x^{1/2}$ is $x^{-1/2}$ and the *GCF* of 12 and 8 is 4 so we have

$$12x^{-1/2} + 8x^{1/2} = 4x^{-1/2}(3 + 4x)$$

Note this answer may also be written as $\dfrac{4(3 + 4x)}{\sqrt{x}}$.

■■■

The above examples illustrate what is called the factoring out of a *monomial* expression. In Example 1,

Monomial Factors

the expression was the constant 3, in Example 2 it was $36x^2$, in Example 3 it was $5x^3y^3z^4$, and in Example 4 it was $4x^{-1/2}$. Monomial factoring occurs in various computations in calculus. The next two examples are typical of those arising in calculus.

Example 5

Factor $20x^5(x^2 + 1)^9 + 4x^3(x^2 + 1)^{10}$.

Solution

Observe that the main difference between this example and the others considered is that one of the factors is the binomial expression $(x^2 + 1)$. We can let $A = (x^2 + 1)$ then the expression becomes

$$20x^5A^9 + 4x^3A^{10}$$

This is now easily factored as

$$20x^5A^9 + 4x^3A^{10} = 4x^3A^9(5x^2 + A)$$

now replacing A with $(x^2 + 1)$ we have,

$$20x^5(x^2 + 1)^9 + 4x^3(x^2 + 1)^{10} = 4x^3(x^2 + 1)^9(5x^2 + (x^2 + 1)) = 4x^3(x^2 + 1)^9(6x^2 + 1)$$

■■■

We remark that letting $A = (x^2 + 1)$ made the problem more recognizable, however, it was not really necessary.

Example 6

Factor $3x^2\sqrt{x^2 + 1} + \dfrac{x^2}{\sqrt{x^2 + 1}}$.

Solution

As in the previous example, we substitute for the binomial term. Let $A = x^2 + 1$, then the expression to be factored is

$$3x^2\sqrt{A} + \frac{x^2}{\sqrt{A}} = 3x^2A^{\frac{1}{2}} + x^2A^{-\frac{1}{2}}$$

The *GCF* is $x^2A^{-\frac{1}{2}}$, so we have

$$3x^2\sqrt{A} + \frac{x^2}{\sqrt{A}} = 3x^2A^{\frac{1}{2}} + x^2A^{-\frac{1}{2}} = x^2A^{-\frac{1}{2}}(3A + 1) = \frac{x^2(3A + 1)}{A^{\frac{1}{2}}}$$

We now replace A with $x^2 + 1$, and we have that

$$3x^2\sqrt{x^2+1} + \frac{x^2}{\sqrt{x^2+1}} = \frac{x^2(3(x^2+1)+1)}{(x^2+1)^{\frac{1}{2}}} = \frac{x^2(3x^2+4)}{(x^2+1)^{\frac{1}{2}}}$$

■■■

We next want to consider examples where the expression to be factored is a trinomial (three terms) and its factoring is the product of two binomials (two terms).

Let us first look at the multiplication of two binomials, for example,

$$(2x + 3)(3x + 4) = 6x^2 + 17x + 12$$

Of course, the expression on the right was obtained by multiplying together the two binomial expressions on the left. Sometimes we call the multiplication process the *FOIL* method which is briefly reviewed on page 35. Use of the *FOIL* method is an effective means by which we may *quickly* multiply binomial factors.

Now we want to reverse the question, given the trinomial term $6x^2 + 17x + 12$ how could we factor it into the product of two binomial terms? We begin as follows:

Factoring Trinomials $6x^2 + 17x + 12 = (Ax + b)(Cx + d)$

The problem reduces to finding the numbers *A, b, C* and *d*. One process by which we obtain these four numbers is trial and error. Observe that we must have that

$$AC = 6$$
$$bd = 12$$

But there is another condition, namely when the multiplication process is performed, we must get 17 as the coefficient of the *x* term. This term results from the sum of the outer and inner terms in the multiplication of the two binomials. The *FOIL* method is most useful in quickly checking this term.

For $AC = 6$ we could try 6 and 1 or 3 and 2. Thus, to start, we write

$$(6x + \ )(x + \ ) \text{ or } (3x + \ )(2x + \ )$$

For $bd = 12$ we could have the factors 12 and 1, 6 and 2 or 4 and 3,

Thus, possible factors are:

$(6x + 12)(x + 1)$	$(3x + 12)(2x + 1)$
$(6x + 1)(x + 12)$	$(3x + 1)(2x + 12)$
$(6x + 6)(x + 2)$	$(3x + 6)(2x + 2)$
$(6x + 2)(x + 6)$	$(3x + 2)(3x + 6)$
$(6x +4)(x + 3)$	$(3x +4)(2x + 3)$
$(6x + 3)(x +4)$	$(3x + 3)(2x +4)$.

At first glance, it appears that there are many cases that need to be tested. Actually, some of these can be eliminated immediately, and we will describe how shortly. Of course, the only case that is correct is the one which gives 17 as the coefficient of x, and we have

$$6x^2 + 17x + 12 = (2x + 3)(3x + 4)$$

We state some preliminary steps which often minimize the number of steps in obtaining the factoring of a trinomial expression.

1.   **Write the terms in descending powers**
2.   **If necessary, rewrite the expression so the coefficient of the highest power is positive**
3.   **Factor out any common monomial terms.**
4.   **Use the *FOIL* method to quickly check the various cases**

Step 3 often eliminates cases. For example, $6x^2 + 17x + 12$ has no common factors, so the choice $(6x + 12)(x + 1)$ would be rejected since the first factor $(6x + 12)$ has 6 as a common factor but the original trinomial does not have this common factor, therefore any possible factor cannot have a common factor. Similarly for expressions containing $(6x + 6)$ or $(6x + 2)$ or $(6x + 4)$ or $(6x + 3)$ or $(3x + 12)$ or $(2x + 12)$ or $(3x + 6)$ or $(2x + 2)$ or $(3x + 3)$, since each of these has a common factor while the original trinomial does not. That leaves only the cases $(6x + 1)(x + 12)$ or $(6x + 3)(x + 4)$ or $(3x + 4)(2x + 3)$ to be tested, using the *FOIL* method, the correct factors are quickly determined.

Some other remarks. Suppose the coefficient of the highest term is positive and the trinomial to factored has one of the following four forms:

$$( )x^2 + ( )x + ( ) \qquad\qquad (1)$$
$$( )x^2 - ( )x + ( ) \qquad\qquad (2)$$
$$( )x^2 + ( )x - ( ) \qquad\qquad (3)$$
$$( )x^2 - ( )x - ( ) \qquad\qquad (4)$$

where ( ) stands for some positive number.

Since form (1) has all positive signs all the signs in the binomial factors will be positive, for example,

$$6x^2 + 17x + 12 = (2x + 3)(3x + 4)$$

Since form (2) has a negative coefficient for the x-term while its constant term is positive; the constants in the binomial factors must both be negative, for example,
$$6x^2 - 17x + 12 = (2x - 3)(3x - 4)$$

When cases (3) or (4) occur, the constants must have opposite signs, for example,

$$6x^2 - x - 12 = (2x - 3)(3x + 4)$$

or

$$6x^2 + x - 12 = (2x + 3)(3x - 4)$$

These observations help reduce the time it takes to factor a trinomial expression.

Example 7
Factor $2x^2 + 13x + 15$.

Solution
This expression is form (1), so all the constants will be positive. The factors of 2 are 2 and 1, the factors of 15 are 15 and 1 or 5 and 3, thus we have the following possibilities:

$$(2x + 5)(x + 3)$$
$$(2x + 3)(x + 5)$$
$$(2x + 1)(x + 15)$$
$$(2x + 15)(x + 1)$$

Multiplication, using the FOIL method, quickly reveals that $2x^2 + 13x + 15 = (2x + 3)(x + 5)$.
■■■

Example 8
Factor $18x^2 - 21x + 6$

Solution
First we observe that we can factor out the common monomial factor 3, thus we have

$$18x^2 - 21x + 6 = 3(6x^2 - 7x + 2)$$

The problem now reduces to factoring the trinomial $6x^2 - 7x + 2$, which has form (2). The factors of 6 are 6 and 1 or 3 and 2; while the factors of 2 are 2 and 1, thus we have the following possibilities:

$$(6x - 1)(x - 2)$$

or

$$(3x - 2)(2x - 1)$$

Note: that $(6x - 2)(x - 1)$ is not a possibility because the first factor has 2 as a common factor and we already factored out all common terms, similarly for $(3x - 1)(2x - 2)$.

Of these two possibilities, $(6x^2 - 7x + 2) = (3x - 2)(2x - 1)$, therefore, we have that

$$18x^2 - 21x + 6 = 3(6x^2 - 7x + 2) = 3(3x - 2)(2x - 1)$$

■■■

Example 9
Factor $15x^2 + 14x - 8$.

Solution
The factors of 15 are 15 and 1 or 5 and 3; factors of 8 are 8 and 1 or 4 and 2. This given trinomial has form (3), thus the constant terms have opposite sign. We have as possibilities,

$(15x - 8)(x + 1)$	$(15x - 1)(x + 8)$
$(15x + 8)(x - 1)$	$(15x - 4)(x + 2)$
$(15x + 1)(x - 8)$	$(15x + 4)(x - 2)$

$(15x + 2)(x - 4)$ $(5x - 1)(3x + 8)$
$(15x - 2)(x + 4)$ $(5x - 4)(3x + 2)$
$(5x - 8)(3x + 1)$ $(5x + 4)(3x - 2)$
$(5x + 8)(3x - 1)$ $(5x + 2)(3x - 4)$
$(5x + 1)(3x - 8)$ $(5x - 2)(3x + 4)$

Checking the multiplication using the *FOIL* method gives

$$15x^2 + 14x - 8 = (5x - 2)(3x + 4)$$

■■■

Note that

$$(5x + 2)(3x - 4) = 15x^2 - 14x - 8$$

Switching of the signs on the constants reverses the sign of the coefficient of the x-term. The next example illustrates how this reversal is used when the sign of the coefficient of the x-term is opposite of what it should be.

Exercise 10
Factor $20x^2 + 7x - 6$.

Solution
6 has two possible factors, 6 and 1 or 3 and 2, and since the sign of 6 is negative, the factors have to be opposite. To start, our possibilities are (-6)($+1$) or (-3)($+2$) where the blanks are the x terms that need to be determined. Note, there are two more cases, with the signs switched, but we can fix up the signs later as needed.

Suppose we guess and first try the case (-3)($+2$). We next have to consider the factors of 20, 20 and 1, or 10 and 2 or 5 and 4. Let us try 5 and 4.
$$(4x - 3)(5x + 2)$$

(Note the $4x$ term cannot be with the 2 as they would then have a common factor which the original trinomial does not have.)

Multiplying out we have $20x^2 - 7x - 6$, we are off on the sign of the coefficient of x. Therefore, we need only reverse the signs and we have,
$$20x^2 + 7x - 6 = (4x + 3)(5x - 2)$$

■■■

It is an easy matter to deal with trinomials whose leading term is negative, just factor out the negative sign as the next example illustrates.

Example 11
Factor $-6w^2 - w + 1$.

Solution

We first factor out the negative sign giving us

$$-(6w^2 + w - 1)$$

We now factor $6w^2 + w - 1 = (2w + 1)(3w - 1)$, therefore we have

$$-6w^2 - w + 1 = -(2w + 1)(3w - 1)$$

■■■

(Note the previous answer could have been written as $(1 + 2w)(1 - 3w)$ why?)

Exercise 12

Factor $-36x^3 + 51x^2 - 18x$

Solution

We first observe that the leading coefficient is negative as well as their being a common monomial factor. We therefore factor out a negative monomial and we have

$$-36x^3 + 51x^2 - 18x = -3x(12x^2 - 17x + 6)$$

The problem reduces to factoring $12x^2 - 17x + 6$, we have (verify!)

$$12x^2 - 17x + 6 = (4x - 3)(3x - 2)$$

therefore,

$$-36x^3 + 51x^2 - 18x = -3x(12x^2 - 17x + 6) = -3x(4x - 3)(3x - 2)$$

■■■

Now that we have reviewed the factoring of the trinomial, we illustrate, through the next few examples, how we may apply the procedure to other trinomial forms.

Example 13

Factor (a) $10x^2 - 23xy + 12y^2$ (b) $10x^2y^2 - 23xy + 12$.

Solution

The factoring in this problem is almost the same as the factoring of

$$10x^2 - 23x + 12 = (2x - 3)((5x - 4) \text{ (verify!)}$$

The only difference between this expression and those appearing in the problem is the placement of the y-terms. Therefore, we need only place the y-terms appropriately in the above factoring to obtain the required factoring.

(a) The given expression has y appearing in the second term and squared in the third term. We can therefore factor it as follows.

$$10x^2 - 23xy + 12y^2 = (2x - 3y)((5x - 4y)$$

Note that the placement of the y-term in each factor yields the correct result. Multiply out to verify the correctness of the factoring.

(b) The given expression has the y-squared term next to the x-squared term, we can accomplish this with the factoring

$$10x^2y^2 - 23xy + 12 = (2xy - 3)((5xy - 4)$$

■■■

Example 14
Factor $25x^2 + 30x + 9$.

Solution
If we factor, we have

$$25x^2 + 30x + 9 = (5x + 3)(5x + 3) = (5x + 3)^2$$

■■■

Note in the previous example, the factoring gave us the product of a binomial with itself, a prefect square. Thus, the given expression is called a *perfect square*. We could write down an expression which is a general formulation of a perfect square but it is easy enough to recognize one from its factoring.

Perfect Squares

The product of two binomial factors always results in a trinomial with a "middle x-term," with one exceptional case. Observe that during the multiplication process, the x-term arises from the "inner" and "outer" multiplications. If these terms are equal and opposite, then they cancel. Consider the following multiplication:

outer
$\updownarrow$
$$(2x - 3)(2x + 3) = 2x^2 - 6x + 6x - 9 = 2x^2 - 9$$
$\updownarrow$
inner

Conjugates - The Difference of Two Squares

Notice that the binomial terms $(2x - 3)$ and $(2x + 3)$ are identical except for their connecting signs; such terms are called *conjugates* of each other. More generally, the factors $(ax + b)$ and $(ax - b)$ are said to be conjugates of each other and their product is $a^2x^2 - b^2$. This last expression is sometimes called the difference of two squares because the coefficient of x^2 and the constant terms are both squares.

Example 15
Factor $16x^2 - 25$.

Solution
since 16 and 25 are perfect squares, 16 is the square of 4 and 25 is the square of 5, we have

$$16x^2 - 25 = (4x - 5)(4x + 5)$$

■■■

We observed when we studied radicals and their conjugates (Section 4) that the terms do not have to be perfect squares, that is, we are able to factor any expression of the form $ax^2 - b$, if we allow irrational numbers. Moreover, we extended this idea to complex conjugates (Section 5), allowing factoring when b is negative.

Factoring by Grouping

Another type of factoring that is sometimes used is the so called factoring by grouping. We illustrate with a few examples, and then show how it is gives an alternative way of factoring trinomials.

Example 16
Factor $9x + 6 + 3ax + 2a$

Solution
Observe that if we group together and factor the first two terms and the last two terms, we have

$$(9x + 6) + (3ax + 2a) = 3(3x + 2) + a(3x + 2)$$

Now observe the common factor $(3x + 2)$, so we have

$$9x + 6 + 3ax + 2a = 3(3x + 2) + a(3x + 2) = (3x + 2)(3 + a)$$

■■■

It turns out that using factoring by groupings is an alternative way of factoring trinomials. Consider the trinomial
$$3x^2 - 7x - 6$$

Suppose we look for integers whose sum is -7 (the coefficient of the x-term) and whose product is $3 \cdot (-6)$ = 18 (the product of the coefficient of the x^2 term and the constant term). Inspection quickly yields the two integers -9 and 2. We rewrite the x-terms of the expression as follows:

$$3x^2 - 7x - 6 = 3x^2 - 9x + 2x - 6$$

We now factor this expression by grouping.

$$3x^2 - 7x - 6 = 3x^2 - 9x + 2x - 6 = 3x(x - 3) + 2(x - 3) = (x - 3)(3x + 2)$$

Of course, you will get the same factoring if you proceed directly by trial and error. This alternative method illustrates an important application of factoring by grouping.

Example 17
Using factoring by grouping, factor $6x^2 + 17x + 12$.

Solution
We need to find two integers whose sum is 17 and whose product is $6 \cdot 12 = 72$; 9 and 8 are the required integers, so we have

$$6x^2 + 17x + 12 = 6x^2 + 9x + 8x + 12 = 3x(2x + 3) + 4(2x + 3) = (2x + 3)(3x + 4)$$

■■■

Note, it does not make any difference how we write the sum of the x-terms. For example if we reverse their order and write

$$6x^2 + 17x + 12 = 6x^2 + 8x + 9x + 12 = 2x(3x + 4) + 3(3x + 4) = (3x + 4)(2x + 3)$$

You recall that a quadratic equation is any equation of the form $ax^2 + bx + c = 0$ with $a \neq 0$. When the trinomial on the left hand side of the equation is easily factored then we may solve the quadratic equation by factoring. The solution follows from the principle that if $AB = 0$ then either $A = 0$ or $B = 0$. We illustrate with an example.

Quadratic Equations and Factoring

Example 18
Solve the quadratic equation $6x^2 + 17x + 12 = 0$.

Solution
Factoring, we have
$$6x^2 + 17x + 12 = (3x + 4)(2x + 3) = 0$$

Therefore we set each factor to zero to find the two possible solutions for x, namely,

$$3x + 4 = 0 \text{ or } 2x + 3 = 0$$

The first equation yields $x = -4/3$ and the second $x = -3/2$. Thus the two solutions are $x = -3/2$ or $-4/3$.
■■■

Example 19
Solve the quadratic equation $4x^2 - 7x = 0$.

Solution
Factoring, we have
$$4x^2 - 7x = = x(4x - 7) = 0$$

Thus, we have either $x = 0$ or $4x - 7 = 0$, giving the two solutions, $x = 0$ or $x = 7/4$.
■■■

We emphasize, that the quadratic must equal to zero to use this approach. Consider the following example.

Example 20
Solve the equation $(3x - 4)(2x + 5) = -7$

Solution
We first multiply out to obtain
$$(3x - 4)(2x + 5) = 6x^2 + 7x - 20 = -7$$

or

$$6x^2 + 7x - 13 = 0$$

$$6x^2 + 7x - 13 = (6x + 13)(x - 1) = 0$$

Therefore

$$6x + 13 = 0 \text{ or } x - 1 = 0$$

Solving each of these equations for x yields the two solutions, $x = -13/6$ or 1.
■■■

The next example illustrates that the principle $AB = 0$ implies either $A = 0$ or $B = 0$ may be extended to any number in factors, in particular, if $ABC = 0$, then $A = 0$ or $B = 0$ or $C = 0$.

Exercise 21
Solve the equation $20x^3 + 14x^2 = 24x$.

Solution
We write the expression as $20x^3 + 14x^2 - 24x = 0$ and factor to obtain

$$20x^3 + 14x^2 - 24x = 2x(10x^2 + 7x - 12x) = 2x(5x - 4)(2x + 3) = 0$$

This product will be zero when either $2x = 0$ or $5x - 4 = 0$, or $2x + 3 = 0$. Solving each of these equations for x yields $x = 0$ or $x = 4/5$ or $x = -3/2$.
■■■

Exercise set 7
Completely factor each of the following

1. $10x + 25y$

2. $125a + 75b$

3. $12a - 16b$

4. $18x + 27y$

5. $ab + bc$

6. $Yb - Y$

7. $xa + a$

8. $120x^2y + 192x^3y^2$

9. $60a^3b^2 + 120a^2b^3$

10. $90m^3n^2p^4 - 150m^4n^3p^3$

11. $96x^3y^2z^5 + 160x^2y^4z^6$

12. $4x^{1/2} + 3x^{3/2}$

13. $15y^{3/2} + 20y^{5/2}$

14. $12x^{-1/2} + 18x^{3/2}$

15. $20x^{-3/2} + 36x^{1/2}$

16. $2x(x^2 + 1)^2 + 4x(1 - x^2)(x^2 + 1)$

17. $28x^6(2x^2 + 3)^6 + 5x^4(2x^2 + 3)^7$

18. $8x^{10}(2x^2 + 7)^{15} + 80x^9(2x^2 + 7)^{16}$

19. $3x^5(x^2 + 2x + 3)^7 + 15x^4(x^2 + 2x + 3)^8$

20. $6x^4(x^2 + 1)^{1/2} + 12x^2(x^2 + 1)^{3/2}$

21. $\dfrac{3x^6}{(x^3 + 1)^{2/3}} + 12x^3(x^3 + 1)^{1/3}$

22. $32x^3(3x^2+2)^{1/4} + \dfrac{12x^5}{(3x^2+2)^{3/4}}$

23. $2x^2 - 3x - 2$

24. $10x^2 + 9x + 2$

25. $3x^2 + 4x - 4$

26. $6x^2 + 13x + 6$

27. $6x^2 - 5x - 6$

28. $6x^2 - 13xy + 6y^2$

29. $12x^2 + 7x - 10$

30. $12x^2 - 23xy + 10y^2$

31. $18x^2 + 39x + 20$

32. $18x^2y^2 + 9xy - 20$

33. $24x^2y^2 - 73xy + 24$

34. $24x^2 - 55x - 24$

35. $24x^2 + 4x - 60$

36. $18x^2 + 33x - 30$

37. $-24x^2 - 32x + 70$

38. $20x^3 + 52x^2 - 24x$

39. $-72x^4 - 12x^3 + 24x^2$

40. $9x^2 + 24xy + 16y^2$

41. $x^2 + 2xy + y^2$

42. $x^2 - 2xy + y^2$

43. $4x^2 + 12xy + 9y^2$

44. $9x^2y^2 - 30xy + 25$

45. $25x^2 - 30x + 9$

46. $x^2 - 4$

47. $4x^2 - 9$

48. $64x^2 - 25y^2$

49. $16x^2 - 49$

In exercises 43 - 53 solve the given equation for x.

50. $x^2 - 7x + 12 = 0$

51. $x^2 - 8x + 12 = 0$

52. $6x^2 - 5x = 6$

53. $-12x^2 - 7x + 10 = 0$

54. $20x^3 + 52x^2 = 24x$

55. $-36x^4 - 6x^3 + 12x^2 = 0$

56. $6x^2 - 14x = 0$

57. $12x^2 = 9x$

58. $25 - 9x^2 = 0$

59. $\dfrac{3}{8}x^2 + \dfrac{1}{2}x - \dfrac{1}{2} = 0$

60. $\dfrac{1}{3}x^2 - \dfrac{1}{2}x = \dfrac{1}{3}$

In exercises 54 - 62, factor the given expression by grouping.

61. $xw + 2x + 3w + 6$

62. $6a + 3ab + 4 + 2b$

63. $xy + 3y + 2x + 6$

64. $ab^2c + 3ab + 2bc + 6$

65. $2x^2 - 3x - 2$

66. $3x^2 + 4x - 4$

67. $6x^2 + 13x + 6$

68. $12x^2 + 7x - 10$

69. $9x^2 + 24x + 16$

70. $25x^2 - 30x + 9$

POSTTEST 7- Time 15 minutes

Each question is worth one point. Completely Factor each expression

1. $25x^3y^4 + 30x^4y^2$

2. $15x^{\frac{1}{4}} + 20x^{\frac{5}{4}}$

3. $18x^2(x^2 + 3)^{12} + 27x^4(x^2 + 3)^{11}$

4. $3x\sqrt{2x^2 + 1} - \dfrac{6x^3}{\sqrt{2x^2 + 1}}$

5. $24x^2 + 2x - 15$

6. $18x^2 - 39xy + 20y^2$

7. $-15 + 38x - 24x^2$

8. $9x^2 - 49$

9. $25x^2 - 81y^2z^2$

10. $4ax^2 - 16a + 8 - 2x^2$

(Notes)

8. Solving Linear Equations

» **Addition and Multiplication Properties**
» **Linear Equations with Fractions**
» **Linear Equations with Decimals**
» **Solving for a Particular Variable**
» **Applications of Linear Equations**

PRETEST 8 - Time 10 minutes

Each question is worth one point. Solve for the unknown:

1. $x + 2 = 3$

2. $x + 5 = 2$

3. $3y - 2 = 7$

4. $-2x + 4 = -10$

5. $2z - 5 = 5z - 3$

6. $0.2x + 3.2(2 - 5x) = .5$

7. $\frac{3}{5}x = 12$

8. $\frac{3}{4}w - \frac{1}{4} = \frac{2}{3}w + \frac{7}{2}$

9. $\frac{2x}{3x - 2} = \frac{4}{3}$

10. Solve for y: $2x - 3y = 5$

In this section we review the method of solving a linear equation. The objective of this section is to remind you how to solve an equation of the form

$$ax + b = c$$

for x (we assume, of course, that $a \neq 0$). The basic idea of solving an equation is to *isolate* the unknown on one side of the equation. The solution is then the number on the other side. Usually, two steps are required. First isolate the term containing the unknown then isolate the unknown itself. This often takes the application of two properties: the first is the *addition property* which states that the same term may be added (or subtracted) to each side of an equation. For example if $A + D = E$ then $A + D - D = E - D$ or

Addition and Multiplication Properties

$A = E - D$. This property may be rephrased as follows: an expression may be moved from once side of an equation to the other provided its sign is reversed. Notice we move D from the left side of the equation $A + D = E$ to the right side by changing its sign and writing $A = E - D$. Sometimes the procedure is called *transposition*. Once the unknown term is isolated, we use the *multiplication property* to solve for it; this property states that two sides of an equation may be multiplied by the same (non-zero) expression. For example, if $A = B$ then $AC = BC$ or if $AB = D$, then if we multiply this equation by $1/B$ (or equivalently divide by B) we have that $A = D/B$.

We illustrate how these properties are used in the following examples.

Example 1
Solve the following equation for x: $x + 2 = 5$.

Solution
We isolate the x term by transposing the 2 to the right hand side of the equation and write $x = 5 - 2$, or $x = 3$.

It is important to check the result. This is performed by substituting our result for x wherever it appears in the original problem. In this example,

$$3 + 2 \overset{?}{=} 5$$
$$5 = 5$$

Thus, our solution checks the original equation

■■■

Example 2
Solve the following equation for x: $3x - 2 = 7$

Solution
We isolate the x term by transposing the -2 to the right hand side of the equation and write

$$3x = 7 + 2$$

or

$$3x = 9$$

we next solve for x by multiplying both sides of the equation by the reciprocal of 3, 1/3 yielding,

$$(1/3)3x = (1/3)9$$

or

$$x = 3$$

We check our result.

$$3(3) - 2 \overset{?}{=} 7$$
$$9 - 2 \overset{?}{=} 7$$
$$7 = 7$$

which checks our solution.

■■■

Note: in the last example, instead of multiplying each term on both sides of the equation by 1/3, we could have equivalently divided each term by 3.

Example 3

Solve the following equation for x: $\frac{2}{3}x = 5$

Solution
Method I
Since the x term is already isolated, we need only multiply both sides of the equation by the reciprocal of 2/3, which is 3/2. Thus,

**Linear Equations
with Fractions**

$$\frac{2}{3}x = 5$$

$$\left(\frac{3}{2}\right)\frac{2}{3}x = \left(\frac{3}{2}\right)5$$

$$x = \frac{15}{2}$$

Method II
Recall that whenever an equation contains fraction, we may multiply every term on each side of the equation by the *LCD* (lowest common denominator) to clear the fractions. In this example, we need only multiply each term on either side of the equation by the *LCD* which is 3 and then complete the solution as follows.

$$\frac{2}{3}x = 5$$

$$3 \cdot \frac{2}{3}x = 3 \cdot 5$$

$$2x = 15$$

$$(\frac{1}{2})2x = (\frac{1}{2})15$$

$$x = \frac{15}{2}$$

We leave the checking of the solution in this and the rest of the examples as exercises for you.
■■■

As we see from the last example, there is often more than one way to solve an equation. Sometimes one method may be preferred to another as it results in less work in obtaining the solution.

Example 4
Solve the equation $5y - 7 = 2y - 11$ for y.

Solution
We need to isolate the y term. First we transpose the -7 from the left to the right side of the equation by changing its sign, so we have

$$5y = 2y - 11 + 7$$

$$5y = 2y - 4$$

We next transpose the $2y$ term to the left hand side by changing its sign to obtain

$$5y - 2y = 4$$

or

$$3y = 4$$

and finally, we multiply both sides by 1/3 (or equivalently divide both sides by 3) to obtain

$$y = 4/3$$

■■■

Example 5
Solve the equation $\dfrac{2}{3}w + \dfrac{3}{4} = -\dfrac{1}{6}w + \dfrac{5}{2}$ for w.

Solution
We multiply every term on both sides of the equation by the *LCD* which is 12 to obtain

$$12 \cdot \frac{2}{3}w + 12 \cdot \frac{3}{4} = 12 \cdot \left(-\frac{1}{6}w \right) + 12 \cdot \frac{5}{2}$$

$$8w + 9 = -2w + 30$$

$$8w = -2w + 30 - 9$$

$$8w + 2w = 21$$

$$10w = 21$$

$$w = 21/10$$

'■■■

Example 6
Solve the equation $0.5x + 0.2(3 - 2x) = 0.04$

Solution

Linear Equations with Decimals

Just as with fractions, where they are cleared by multiplying by the *LCD*, we may do the same thing with decimals. Observe that the smallest decimal involves hundredths (0.04), so we multiply each term on each side of the equation by 100 to obtain

$$100 \cdot 0.5x + 100 \cdot 0.2(3 - 2x) = 100 \cdot 0.04$$

$$50x + 20(3 - 2x) = 4$$

$$50x + 60 - 40x = 4$$

$$10x = -56$$

$$x = -\frac{56}{10} = -5.6$$

(Note that we could have written the answer as -28/5, but since the original problem involved decimals, we left the answer in decimal form.)

■■■

Example 7
Solve the equation
$3(2m - 4) + 3(5 - 6m) = 9 - 4(7 - 4m)$ for m.

Solution
We first distribute, combine like terms, and transpose to isolate m and then solve.

$$6m - 12 + 15 - 18m = 9 - 28 + 16m$$

$$-12m + 3 = -19 + 16m$$

$$-12m = -19 + 16m - 3$$

$$-12m = -22 + 16m$$

$$-12m - 16m = -22$$

$$-28m = -22$$

$$m = 22/28 = 11/14$$

■■■

Sometimes, more than one variable may appear in an equation and we are asked to solve for one of them in terms of the others. Our tactics are still the same, isolate the term containing the desired variable.

Solving for a Particular Variable

Example 8
Solve the equation $3x + 4y = 24$ for (a) x (b) y.

Solution
(a) As before we isolate the x term by transposing the y term to obtain

$$3x = 24 - 4y$$

Now we divide each side of the equation by 3, to obtain

$$x = \frac{24 - 4y}{3}$$

or

$$x = 8 - \frac{4}{3}y.$$

(b) We next solve the original equation for y. We isolate the y term by transposing the x term to the right hand side of the equation to obtain

$$4y = 24 - 3x$$

Dividing each side of the equation by 4 gives

$$y = \frac{24 - 3x}{4} = 6 - \frac{3}{4}x$$

often, this last equation is written as

$$y = -\frac{3}{4}x + 6$$

■■■

It is important that we be able to use mathematics as means by which we may solve applied problems. The next few examples illustrate how we translate an applied problem into a mathematical equation and then solve the equation.

Applications of Linear Equations

Example 9
A 24 foot rope is cut into two pieces. The longer piece is 3 feet longer than twice the shorter piece. Find the length of the longer piece.

Solution
Let $x =$ the length of the shorter piece.

Let us translate the statement *the longer piece is 3 feet longer than twice the shorter piece.*

The longer piece $= 3 + 2x$

(Note that *is* translates into =, *more than* into +.)

The sum of the shorter piece and the longer piece is the total length which is 24 feet (see Figure 1), therefore, we have

$$x + (3 + 2x) = 24$$

$$3x + 3 = 24$$

$$3x = 24 - 3$$

$$3x = 21$$

$$x = 7$$

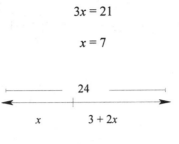

Figure 1

Thus, the shorter piece is 7 feet long and the longer piece is $3 + 2(7) = 17$ feet long.

■■■

Example 10
The sum of Susan and Mark's present ages is 27 years. Ten years from now, twice Susan's age (then) will exceed Mark's age (then) by 28 years. What are their present ages?

Solution
Let Mark's present age $= m$, since the sum of their present ages is 27, we have
Susan's present age $= 27 - m$.

In 10 years their ages will be $(m + 10)$ and $(27 - m + 10)$ respectively. At that time, twice Susan's age (two times her age) will be 28 years more than Mark, so we must have

$$2(27 - m + 10) = (m + 10) + 28$$
twice Susan's age is 28 years more that Mark's age

Note that twice Susan's age exceeds Marks age by 28, so for them to be equal, we need to add 28 to Mark. Solving, we have

$$2(37 - m) = m + 38$$

$$74 - 2m = m + 38$$

$$-3m = -36$$

$$m = 12$$

Thus, Mark's present age is 12 years, and Susan's is $27 - 12 = 15$ years.

■■■

Exercise 11
The Smiths want to add a rectangular porch to their home. Due to the remaining lot size, the length of the proposed porch must be 7 feet more than twice the width. If the perimeter of the room is to be 50 feet, determine the dimensions of the room.

Solution

We recall that the perimeter of a rectangle of length l and width w is $P = 2l + 2w$.

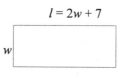

$$l = 2w + 7$$

Figure 2

We labeled the above rectangle, representing the porch, in Figure 2 with all the pertinent information. From the perimeter relationship, we have that

$$50 = 2l + 2w = 2(2w + 7) + 2w$$

or

$$50 = 4w + 14 + 2w$$

$$36 = 6w$$

$$6 = w$$

Thus, the width of the porch is to be 6 feet, and the length $2(6) + 7 = 19$ feet.

■■■

Recall that when objects (cars, planes, bicycles, people, etc.) are moving at a constant rate (speed), the relationship between the *rate* (r), the *distance* (d) and the *time* (t) is given by the formula

$$rt = d$$

(that is, *rate* times *time* is equal to *distance*.) If we solve this equation for r we have that $r = d/t$, or if solve this equation for t, we have that $t = d/r$. Depending on what needs to be solved for, we use one of these equivalent forms to solve motion problems. The next example illustrates a simple application of this result.

Example 12

John and Isabel leave the parking lot and travel in opposite directions. If John averages 50 mph and Isabel averages 60 mph, how long does it take them to be 385 miles apart?

Solution

Let t be the time they need to travel to be 385 miles apart. Since John's average rate is 50 mph, the distance he travels in this time is $50t$. Similarly, the distance Isabel travels in this time is $60t$.

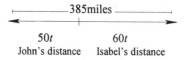

Figure 3

Note that the sum of John and Isabel's distances must be 385 miles (see Figure 3), so we have the equation

$$50t + 60t = 385$$

or

$$110t = 385$$

$$t = 385/110 = 3\tfrac{1}{2} \text{ hours}$$

Thus, in 3½ hours they will be 385 miles apart.

■■■

We conclude this section with one more application which further reviews the concept of solving an equation using the method of *cross multiplication*. Recall that if two fractions are equal, that is, if

$$\frac{A}{B} = \frac{C}{D}$$

Then by multiplying each side of the equation by the common denominator BD, it follows that

$$AD = BC$$

Example 13

The denominator of a fraction is one more than twice the numerator. If the numerator is doubled and the denominator is increased by two the resulting fraction is 2/3. Determine the original fraction.

Solution

Let x be the numerator of the original fraction, then the denominator is $1 + 2x$. Therefore the original fraction is represented by $\dfrac{x}{2x+1}$. If the numerator is doubled and the denominator is increased by two, then the new fraction is $\dfrac{2x}{2x+3}$. Since this new fraction is 2/3, we have

$$\frac{2x}{2x+3} = \frac{2}{3}$$

cross multiplying, we have

$$3(2x) = 2(2x+3)$$

or

$$6x = 4x + 6$$

or

$$2x = 6$$

or

$$x = 3$$

Therefore, the original fraction was

$$3/(2(3) + 1) = 3/7$$

■■■

Exercise set 8

In Exercise 1 - 21, solve for the unknown, and then check your solution.

1. $x - 7 = 9$

2. $x + 11 = 5$

3. $5x = 0$

4. $2w - 7 = 5w - 7$

5. $4w - 3 = w + 5$

6. $7m - 4 = 3m + 12$

7. $9 - 5r = 4r + 11$

8. $3(4w - 5) = 7(9 - 3w)$

9. $3p + 2(5p - 6) = 19 + 4(7 - 2p)$

10. $2 + 3(5t - 9) = 7t - 3(9 - 3t)$

11. $5n - 3 = 2(9n - 7) + 14 - 6n$

12. $13 - 5x + 3(2x - 9) = 12x$

13. $\frac{2}{3}x = 8$

14. $\frac{3}{5}y = 9$

15. $\frac{5}{2}z = \frac{4}{3}$

16. $\frac{3}{4}w + 2 = w - \frac{1}{2}$

17. $\frac{3}{4}w - \frac{2}{3}w + \frac{1}{4} = \frac{1}{6}w + 3$

18. $\frac{3}{5}y - \frac{3}{4} = \frac{1}{4}y + \frac{7}{10}$

19. $0.2x - 0.3(5 - 2x) = 0.4$

20. $0.35 + 0.24x = 0.2(5 - 3x)$

21. $0.4r - 0.52(40 - 2r) = 100$

In exercises 22- 29, solve for the indicated variable.

22. For x: $5x - 2y = 19$

23. For y: $5x - 2y = 19$

24. For x: $ax + by = c$

25. For y: $ax + by = c$

26. For C: $F = 9/5C + 32$

27. For h: $V = \pi r^2 h$

28. For b: $A = \frac{1}{2}h(a + b)$

29. For S: $1/R = 1/S + 1/W$

30. Today, Maria's age is 5 years less than twice Louie's age. In five years, the sum of their ages then will be 41. What are their present ages?

31. Mike is 5 years older then Joan. In ten years, the sum of their ages will be 57. What are their present ages?

32. Alphonse took four examinations and his average on for these four exams was 83. If his grades on the first three exams were 82, 88 and 78, what grade did he get on the fourth exam?

33. The length of a rectangle is two feet less than three times its width. If the perimeter is 36 feet, what are the dimensions of the rectangle?

34. An isosceles triangle has a perimeter of 33 centimeters. If the equal sides are 6 centimeters less than twice the unequal side, what is the length of the two equal sides?

35. The sum of three consecutive integers is 96, find the three integers.

36. The sum of two consecutive odd integers is 15 more than the next odd integer. Find the three integers.

37. Boston and New York are approximately 280 miles apart. One train leaves Boston traveling towards New York at an average speed of 80 mph. Another train leaves New York at the same time traveling toward Boston at an average speed of 60 mph. (a) How long will it take them to meet. (b) How far has each train traveled when they meet?

38. Two cars leave a city traveling in the same direction. One travels at 60mph and the other at 45 mph. How long before the cars are 60 miles apart?

39. Two planes leave JFK airport at the same time, one flying north at 500 mph and the other south at 600 mph. In how many hours will they be 4400 miles apart?

40. Two angles are said to be complementary if the sum of the measures of their angles is 90 degrees. If one angle is 15 degrees more than twice the other, what is the measure of the smaller angle?

41. Two angles are said to be supplementary if the sum of the measures of their angles is 180 degrees. If one angle is 30 degrees less than six times the other, what is the measure of the larger angle?

42. James has a collection of nickels and dimes which total to $5.15. If he has 5fewer nickels than dimes, how many dimes has he?

43. The denominator of a fraction is 1 less than three times the numerator. If the numerator is increased by 1and the denominator is decreased by 2 the resulting fraction is ½. Determine the original fraction.

44. The numerator of a fraction is three less than the denominator. Tripling the numerator and quadrupling the denominator results in a fraction equal to 3/5. Determine the original fraction.

POSTTEST 8 - Time 10 Minutes

Each question is worth one point. Solve for the unknown:

1. $3 - x = 2$

2. $5x - 2 = 8$

3. $5 - 7y = 26$

4. $\frac{4}{3}x = 20$

5. $3(2z - 5) - 4(2 - 3z) = 5(9 + 2z)$

6. $\frac{1}{6}x - \frac{2}{3} = \frac{3}{4}x + 2$

7. $\frac{3x}{2 - 5x} = \frac{3}{4}$

8. $8.2(3 + 2x) - 1.1(5x - 4) = 2$

9. Solve for y: $3x - 4y = 9$

10. A nine foot rope is cut into two pieces so that one piece is twice the other. How long is the larger piece?

9. Solving Equations of the Form $ax^2 + b = 0$

» **Isolation of Squared Term**
» **Isolation of Squared Binomial Term**

PRETEST 9- Time 10 minutes

Solve each of the following quadratic equations for the unknown.

1. $4x^2 = 5x$

2. $5x^2 - 20x = 0$

3. $9x^2 - 49 = 0$

4. $16x^2 - 12 = 0$

5. $(2x - 1)(3x + 2) = 24$

6. $4x^2 + 32 = 0$

7. $\frac{3}{4}x^2 - \frac{2}{3} = 0$

8. $(x - 5)^2 - 18 = 0$

9. $3(x - 5)^2 + 81 = 0$

10. $\frac{3}{4}(2x - 5)^2 - 14 = 1$

You have already reviewed in Sections 4, 5 and 7 how to solve quadratic equations by factoring. Unfortunately, not every quadratic equation has factors which are rational numbers, so the factors are not always obvious. Therefore, we shall examine other techniques for solving quadratic equations which will easily yield their solutions, also called their roots.

Consider the equation $x^2 - 9 = 0$. This problem is most easily solved by factoring. We write,

$$(x - 3)(x + 3) = 0$$

Isolation of Squared Term

and we immediately find that $x = -3$ or $x = 3$, or more succinctly as $x = \pm 3$. We can also solve this equation by isolating the x^2 term. Consider the following. Rewrite the equation with the x^2 term isolated on one side of the equation.

$$x^2 = 9$$

We now take the square root of each side, remembering that x could also be negative, so we have

$$x_1 = \sqrt{9} = 3 \text{ or } x_2 = -\sqrt{9} = 3.$$

or written more succinctly as

89

$$x = \pm\sqrt{9} = \pm 3$$

More generally, given the equation $ax^2 + b = 0$, we can write

$$ax^2 = -b$$
$$x^2 = -\frac{b}{a}$$
$$x = \pm\sqrt{-\frac{b}{a}}$$

Of course, we would have to simplify the radical expressions, and sometimes, the solutions could be complex numbers. We illustrate the various cases in the examples that follow.

One general remark: for simplicity in carrying out the algebraic manipulations, the first thing we should do is to clear any fractions. That is, if fractions appear in the quadratic equation, first multiply *every* term on each side of the equation by the *LCD*. This results in an equivalent quadratic equation all of whose terms are integers.

Example 1
Solve the quadratic equation $4x^2 - 25 = 0$.

Solution
Isolating the x^2 term, we have

$$4x^2 = 25$$
$$x^2 = \frac{25}{4}$$
$$x = \pm\sqrt{\frac{25}{4}}$$
$$x = \pm\frac{5}{2}$$

∎∎∎

Example 2
Solve the equation $x^2 - 8 = 0$.

Solution
Isolating the x^2 term, we have

$$x^2 = 8$$
$$x = \pm\sqrt{8}$$
$$x = \pm 2\sqrt{2}$$

∎∎∎

Example 3
Solve the equation $\frac{2}{3}x^2 - \frac{1}{4} = 0$

Solution

We begin by first clearing fractions. We multiply every term on each side of the equation by the *LCD* which is 12.

$$12 \cdot \frac{2}{3}x^2 - 12 \cdot \frac{1}{4} = 12 \cdot 0$$

$$8x^2 - 3 = 0$$

$$8x^2 = 3$$

$$x^2 = \frac{3}{8}$$

$$x = \pm\sqrt{\frac{3}{8}} = \pm\sqrt{\frac{3}{8} \cdot \frac{2}{2}} = \pm\frac{\sqrt{6}}{4}$$

■■■

Example 4

Solve the equation $x^2 + 8 = 0$.

Solution

Isolating the x^2 term, we have

$$x^2 = -8$$

$$x = \pm\sqrt{-8} = \pm i\sqrt{8} = \pm i\sqrt{4}\sqrt{2} = \pm 2\sqrt{2}\,i$$

■■■

Note that in the last example, the radicand was negative resulting in imaginary solutions.

Isolation of Squared Binomial Term

A slight variation on the above method of solution occurs when the left hand side is not x^2, but some expression squared. In such a case, we isolate the squared expression and then take the square roots, as illustrated in the next three examples.

Example 5

Solve the quadratic equation $(z - 5)^2 - 4 = 0$.

Solution

We first isolate the squared term by rewriting the expression as

$$(z - 5)^2 = 4$$

We next take the square roots,

$$z - 5 = \pm\sqrt{4}$$

$$z - 5 = \pm 2$$

$$z = 5 \pm 2$$

$$z_1 = 5 - 2 \text{ or } z_2 = 5 + 2$$

$$z_1 = 3 \text{ or } z_2 = 7$$

■■■

We remark that if we actually squared the expression and rewrote the quadratic as $z^2 - 10z + 21 = 0$, we could have easily solved the problem by factoring. However, that could not be done so easily on the next example.

Example 6

Solve the quadratic equation $(4y + 6)^2 - 12 = 0$

Solution.

$$(4y+6)^2 = 12$$
$$4y+6 = \pm\sqrt{12} = \pm 2\sqrt{3}$$
$$4y = -6\pm 2\sqrt{3}$$
$$y = \frac{-6\pm 2\sqrt{3}}{4}$$

We can reduce this expression by either breaking it into two fractions or by factoring. If we break it into two fractions, we have

$$y = \frac{-6\pm 2\sqrt{3}}{4} = -\frac{6}{4}\pm\frac{2\sqrt{3}}{4} = -\frac{3}{2}\pm\frac{\sqrt{3}}{2} = \frac{-3\pm\sqrt{3}}{2}$$

Alternately, we could factor the expression and write

$$y = \frac{-6\pm 2\sqrt{3}}{4} = \frac{2(-3\pm\sqrt{3})}{4} = \frac{-3\pm\sqrt{3}}{2}$$

Thus, we have the two solutions $y = \dfrac{-3+\sqrt{3}}{2}$ or $\dfrac{-3-\sqrt{3}}{2}$. If we need a numerical approximation, we can use a calculator.

■■■

Try solving this last example by multiplying out the expression and factoring. You will quickly see that it is not a simple matter to find the factors since they are irrational numbers.

Note that the two solutions to the quadratic equation in the previous examples are conjugates of each other. This will be true in general, as we shall see.

Example 7

Solve the quadratic equation $3(4w - 5)^2 + 81 = 0$.

Solution

Isolating the squared term, we have

$$3(4w-5)^2 = -81$$
$$(4w-5)^2 = -27$$
$$4w-5 = \pm\sqrt{-27} = -i\sqrt{27} = \pm 3\sqrt{3}\,i$$
$$4w = 5\pm 3\sqrt{3}\,i$$
$$w = \frac{5\pm 3\sqrt{3}\,i}{4}$$
$$w_1 = \frac{5-3\sqrt{3}\,i}{4} \text{ or } w_2 = \frac{5+3\sqrt{3}\,i}{4}$$

■■■

Once again, we observe the two solutions are conjugates of each other.

We now make one very important observation. Any quadratic equation written in the form

$$(x + B)^2 - C = 0$$

may be solved by isolating the squared binomial term and then take the square roots. It is precisely this observation that will enable us to solve *any* quadratic equation. What will shall do is to learn to rewrite any given quadratic equation in the above form. To do so, we must first learn how to complete the square, the object of the next section.

Exercise set 9
Find the unknown in each of the following exercises.

1. $x^2 = 25$

2. $x^2 = 36$

3. $y^2 - 4 = 0$

4. $y^2 - 49 = 0$

5. $z^2 - 25 = 39$

6. $x^2 + 7 = 16$

7. $x^2 - 4 = 21$

8. $y^2 - 40 = 24$

9. $5x^2 - 7 = 2x^2 + 20$

10. $17x^2 + 20 = 11x^2 + 74$

11. $x^2 = 18$

12. $3z^2 - 24 = 0$

13. $8x^2 - 96 = 0$

14. $2y^2 = 48$

15. $x^2 + 25 = 0$

16. $w^2 = -36$

17. $x^2 + 4 = 0$

18. $y^2 + 49 = 0$

19. $x^2 + 16 = 7$

20. $r^2 + 21 = 5$

21. $x^2 + 20 = 0$

22. $3x^2 + 24 = 0$

23. $6w^2 + 96 = 0$

24. $5y^2 + 12 = 3y^2 - 24$

25. $6y^2 + 9 = 4y^2 - 45$

26. $\frac{2}{3}x^2 - 8 = 0$

27. $\frac{3}{4}y^2 + 24 = 0$

28. $\frac{5}{6}r^2 - \frac{2}{3} = \frac{3}{4}$

29. $\frac{5}{6}w^2 - \frac{1}{4} = \frac{5}{3}$

30. $\frac{1}{8}x^2 + 3 = 0$

31. $(x - 2)^2 = 9$

32. $(x - 3)(x + 2) = 14$

33. $(x + 1)(2x - 3) = 12$

34. $(x + 3)^2 = 25$

35. $(x - 4)^2 + 3 = 39$

36. $(2y - 3)^2 = 49$

37. $(3z - 4)^2 = 64$

38. $(4x - 7)^2 = 18$

39. $(5x + 4)^2 = 20$

40. $(2r - 3)^2 -24 = 0$

41. $(6y - 2)^2 +9 = 41$

42. $(x + 3)^2 = -9$

43. $(x - 4)^2 + 16 = 0$

44. $(y - 5)^2 + 12 = 0$

45. $(3x - 2)^2 + 25 = 0$

46. $(4y + 7)^2 + 36 = 0$

47. $(2y - 3)^2 +24 = 0$

48. $(3x - 5)^2 + 18 = 0$

49. $(6x - 4)^2 + 48 = 0$

50. $\frac{2}{3}(x+3)^2 = 12$

51. $\frac{3}{2}(3y-2)^2 - 18 = 0$

52. $\frac{4}{5}(5w+2)^2 + 25 = 0$

53. $\frac{3}{4}(2z-3)^2 + 24 = 9$

54. $\frac{3}{8}(x+7)^2 + \frac{3}{4} = \frac{15}{2}$

55. $\frac{3}{8}(2r+7)^2 + \frac{15}{2} = \frac{3}{4}$

56. $\frac{1}{5}(3x-5)^2 + \frac{2}{3} = 2$

57. $\frac{1}{5}(2z-5)^2 + 2 = \frac{2}{3}$

POSTTEST 9- Time 10 minutes

Solve each of the following quadratic equations for the unknown.

1. $7x^2 = 3x$

2. $4x^2 - 18x = 0$

3. $16x^2 - 81 = 0$

4. $20x^2 - 45 = 0$

5. $(4x - 5)(4x + 5) = -9$

6. $6x^2 + 120 = 0$

7. $\frac{4}{5}x^2 - \frac{3}{10} = 0$

8. $(x + 3)^2 - 24 = 0$

9. $4(x - 5)^2 + 48 = 0$

10. $\frac{5}{3}(5x+4)^2 - 19 = 1$

10. Completing the Square

» Completion of the Square

PRETEST 10- Time 10 minutes

Each question is worth two points.

1. What number should be added to $x^2 - 8x$ so that it is a perfect square?

2. What number should be added to $x^2 + 3x$ so that it is a perfect square?

Complete the square to solve each of the following quadratic equations. Leave the answer in simplest form.

3. $x^2 - 10x - 3 = 0$ 4. $x^2 + 5x - 5 = 0$ 5. $3x^2 - 12x - 4 = 0$

In the preceding section we observed that any quadratic equation written in the form

$$(x + B)^2 - C = 0 \tag{1}$$

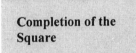

Completion of the Square

may be solved by isolating the squared term and then take square roots. The question that naturally arises is given any quadratic equation, how do we rewrite it so that the unknown appears within a square as in **(1)**. The procedure by which this is accomplished is called *completion of the square*, and in addition to the solution of quadratic equations is useful in numerous applications.

Let us make some observations on several expressions which are identities:

$$x^2 + 6x = (x + 3)^2 - 9$$

The expression may be rewritten as $x^2 + 6x = (x + \frac{1}{2}(6))^2 - (\frac{1}{2}(6))^2$

$$x^2 - 8x = (x - 4)^2 - 16$$

The expression may be rewritten as $x^2 - 8x = (x + \frac{1}{2}(-8))^2 - (\frac{1}{2}(-8))^2$

$$x^2 + 5x = (x + 5/2)^2 - 25/4$$

95

The expression may be rewritten as $x^2 + 5x = (x + \frac{1}{2}(5))^2 - (\frac{1}{2}(5))^2$

Do you see the pattern in each of the examples? We spell it out more generally.

The sum $x^2 + Bx$ may be written as a perfect square as follows:
In words, take one-half the coefficient of the x term, (the sign is part of the coefficient) add it to x, square the sum, then subtract the square of one-half the coefficient of the x term from the previous squared sum.

$$x^2 + Bx = \left(x + \frac{1}{2}B\right)^2 - \left(\frac{1}{2}B\right)^2 \qquad (2)$$

Of course, you may verify that the identity is true by simply multiplying out the right hand side of **(2)**, and comparing it to the left hand side. This identity is useful not only in solving quadratic equations, but in other applications as well, as you shall see later in this text when as well as in calculus.

Example 1
Rewrite the expression $x^2 + 12x$ in the perfect square form as given on the right-hand side of **(2)**.

Solution
The coefficient of x is 12, therefore $\frac{1}{2}(12) = 6$, so we have

$$x^2 + 12x = (x + 6)^2 - (6)^2$$

or

$$x^2 + 12x = (x + 6)^2 - 36$$

■■■

Example 2
Rewrite the expression $x^2 - 10x$ in the perfect square form as given on the right-hand side of **(2)**.

Solution
The coefficient of the x term is -10, one-half this number is -5, so we have

$$x^2 - 10x = (x - 5)^2 - (-5)^2 = (x - 5)^2 - 25$$

■■■

Example 3
Rewrite the expression $x^2 - 3x$ in the perfect square form as given on the right-hand side of **(2)**.

Solution
The coefficient of the x term is -3, one-half this number is -3/2, so we have

$$x^2 - 3x = (x - 3/2)^2 - (-3/2)^2 = (x - 3/2)^2 - 9/4$$

■■■

It is just one more step to see how rewriting a quadratic may be used to solve any quadratic equation. Consider the quadratic equation

$$x^2 - 10x + 10 = 0$$

We isolate the two terms involving x,

$$x^2 - 10x = -10$$

We rewrite the left-hand side of this equation as in Example 2 as $x^2 - 10x = (x - 5)^2 - 25$. The quadratic equation may now be written as

$$(x - 5)^2 - 25 = -10$$

The problem is now in the form we studied in the previous section. We isolate the square term and solve.

$$(x - 5)^2 = 15$$

$$x - 5 = \pm\sqrt{15}$$
$$x = 5 \pm\sqrt{15}$$

■■■

We illustrate how the method of completion of the square is used to solve quadratic equations in the following examples.

Example 4
Solve the quadratic equation $x^2 - 8x + 4 = 0$

Solution
We rewrite the equation as

$$x^2 - 8x = -4$$

and next rewrite the left-hand side as $x^2 - 8x = (x + \frac{1}{2}(-8))^2 - (\frac{1}{2}(-8))^2 = (x - 4)^2 - 16$. Therefore, we have

$$(x - 4)^2 - 16 = -4$$

$$(x - 4)^2 = 12$$

$$x - 4 = \pm\sqrt{12} = \pm 2\sqrt{3}$$
$$x = 4 \pm 2\sqrt{3}$$

■■■

Example 5
Solve the quadratic equation $x^2 - 7x - 6 = 0$

Solution
We rewrite the equation as

$$x^2 - 7x = 6$$

$$x^2 - 7x = (x - 7/2)^2 - (-7/2)^2 = (x - 7/2)^2 - 49/4$$

therefore we have,

$$(x - 7/2)^2 - 49/4 = 6$$

or

$$(x - 7/2)^2 = 49/4 + 6$$

$$(x - 7/2)^2 = 73/4$$

$$x - \frac{7}{2} = \pm\sqrt{\frac{73}{4}} = \pm\frac{\sqrt{73}}{2}$$

$$x = \frac{7}{2} \pm \frac{\sqrt{73}}{2} = \frac{7 \pm \sqrt{73}}{2}$$

■■■

In all the examples considered so far, the coefficient of the x^2 term has been 1. When this is not the case, multiplying each term on both sides of the equation by the reciprocal of the coefficient reduces the problem to an equivalent one with coefficient 1. We illustrate in the following examples.

Example 6
Solve the quadratic equation $4x^2 + 12x - 25 = 0$.

Solution
Multiplying each term on both sides of the equation by the 1/4, the reciprocal of 4, gives

$$x^2 + 3x - 25/4 = 0$$

rewriting, we have

$$x^2 + 3x = 25/4$$

completing the square and solving, we have

$$(x + 3/2)^2 - (3/2)^2 = 25/4$$

$$(x + 3/2)^2 - 9/4 = 25/4$$

$$(x + 3/2)^2 = 25/4 + 9/4$$

$$(x + 3/2)^2 = 34/4$$

$$x + \frac{3}{2} = \pm\sqrt{34/4} = \pm\frac{\sqrt{34}}{2}$$

$$x = -\frac{3}{2} \pm \frac{\sqrt{34}}{2} = \frac{-3 \pm \sqrt{34}}{2}$$

■■■

We remark, that if approximate answers are needed, we would use a calculator to approximate the square roots; for example, in the previous example, the roots to the nearest one-thousandth are 1.416 and -4.416.

Example 7
Solve the quadratic equation $2x^2 - 8x + 25 = 0$

Solution.
We first multiply each term on each side of the equation by the reciprocal of the coefficient of the x^2 term,

½, giving

$$x^2 - 4x + 25/2 = 0$$

or

$$x^2 - 4x = -25/2$$

completing the square, we have

$$(x + ½(-4))^2 - (½(-4))^2 = -25/2$$

or

$$(x - 2)^2 - 4 = -25/2$$

$$(x - 2)^2 = 4 - 25/2$$

$$(x - 2)^2 = -17/2$$

$$x - 2 = \pm\sqrt{-\frac{17}{2}} = \pm i\sqrt{\frac{17}{2}} = \pm i\sqrt{\frac{17}{2} \cdot \frac{2}{2}} = \pm i\sqrt{\frac{17}{4}} = \pm i\frac{\sqrt{34}}{2}$$

$$x = 2 \pm \frac{\sqrt{34}}{2}i = \frac{4 \pm \sqrt{34}}{2}i$$

■■■

Exercise set 10
Complete the square in each of the following by putting in the form $(x + B)^2 - C$.

1. $x^2 + 4x$

2. $x^2 + 6x$

Solve each of the quadratics by first completing the square. When the roots are irrational, also give the solutions to the nearest one-thousandth.

3. $x^2 - 6x$

13. $x^2 + 2x - 10 = 0$

4. $x^2 - 8x$

14. $x^2 + 6x + 3 = 0$

5. $x^2 - 10x$

15. $x^2 - 2x - 10 = 0$

6. $x^2 + 12x$

16. $x^2 - 6x + 3 = 0$

7. $x^2 + 3x$

17. $x^2 + 6x = 3$

8. $x^2 - 3x$

18. $x^2 - 8x = 4$

9. $x^2 - 5x$

19. $x^2 - 12x - 4 = 0$

10. $x^2 + 5x$

20. $x^2 + 5x = 5$

11. $x^2 - 7x$

21. $x^2 - 5x = 5$

12. $x^2 + 9x$

22. $x^2 - 3x - 1 = 0$

23. $x^2 + 3x - 1 = 0$

24. $x^2 + 2x + 10 = 0$

25. $(x - 4)(x + 1) = 6$

26. $(x - 3)(x + 5) = 20$

27. $(2x - 3)(3x + 1) = -4$

28. $(x - 3)(x + 5) = 12$

29. $(x + 3)(x + 2) = -10$

30. $(2x - 3)(3x + 5) = -20$

31. $x^2 - 6x + 23 = 0$

32. $x^2 - 12x = -48$

33. $x^2 + 10x + 50 = 0$

34. $x^2 - 8x + 36 = 0$

35. $x^2 - 4x = 8$

36. $x^2 + 6x + 21 = 0$

37. $x^2 - 3x + 3 = 0$

38. $x^2 + 5x + 10 = 0$

39. $2x^2 + 8x + 1 = 0$

40. $2x^2 + x - 6 = 0$

41. $6x^2 + 5x - 6 = 0$

42. $12x^2 - 7x = 12$

43. $6x^2 + 23x = -20$

44. $2x^2 + 5x + 10 = 0$

45. $4x^2 - 5x + 8 = 0$

46. $\frac{1}{2} x^2 + 5x + 2 = 0$

47. $\frac{1}{2}x^2 - \frac{3}{4}x = \frac{5}{6}$

48. $\frac{3}{5}x^2 - \frac{3}{2}x = \frac{7}{10}$

49. $3 - 4x^2 = 8x$

50. $9 - 3x^2 + 5x = 8 + 2x^2$

51. The area of a triangle is 20 square feet. If the height is three feet less than the base, find the length of the height and base of the triangle.

52. A rock is thrown down from the ledge of a mountain 200 feet above the ground with an initial velocity of 48 feet per second. If the height of the rock, h is given by the equation $h = -16t^2 - 48t + 200$, where t is the time in seconds, how long does it take for the rock to hit the ground? Give your answer to the nearest one-hundredth of a second.

53. A farmer wants to set aside a rectangular plot of land to contain 100 square meters. If the width of the plot is 10 feet less than the length, find the dimensions of the plot. Give your answers rounded to the nearest tenth of a meter.

POSTTEST 10- Time 10 minutes

Each question is worth two points.

1. What number should be added to $x^2 - 12x$ so that it is a perfect square?

2. What number should be added to $x^2 + 5x$ so that it is a perfect square?

Complete the square to solve each of the following quadratic equations. Leave the answer in simplest form.

3. $x^2 - 12x - 4 = 0$ 4. $x^2 + 3x - 9 = 0$ 5. $4x^2 - 16x - 8 = 0$

(Notes)

11. The Quadratic Formula and Applications

» **Quadratic Formula**
» **Clearing Fractions**
» **Applications**
» **Equations Reducible to Quadratics**

PRETEST 11- Time 10 minutes

Solve each of the following using the quadratic formula. Leave answers in simplest radical form. Each question is worth two points.

1. $2x^2 - 4x - 5 = 0$ 2. $4x - x^2 = -1$ 3. $\frac{3}{4}x^2 = \frac{2}{3}x + \frac{1}{6}$ 4. $3x^2 - 6x + 4 = 0$

5. The vertical height h of a rocket measured in feet at time t measured in seconds is given by the equation $h = -16t^2 + 3200t$. To the nearest thousandth of a second, how long does it take the rocket to reach a height of 5,280 feet?

We saw in the last section that any quadratic equation could be solved by the method of completion of the square. Suppose we apply this method to the general quadratic equation

$$ax^2 + bx + c = 0 \tag{1}$$

What should then happen is that our solution should depend on a, b and c. That means we will have a formula for the solutions to *any* quadratic equation. We proceed by solving (1) using completion of the square. We assume that (1) is indeed a quadratic, which means that $a \neq 0$. We multiply every term on each side of (1) by $1/a$, obtaining

$$x^2 + \frac{b}{a}x + \frac{c}{a} = 0$$

we rewrite this equation as

$$x^2 + \frac{b}{a}x = -\frac{c}{a}$$

we next complete the square and obtain

$$\left(x + \frac{b}{2a}\right)^2 - \left(\frac{b}{2a}\right)^2 = \frac{c}{a}$$

103

Transposing, squaring and combining fractions, we have

$$\left(x+\frac{b}{2a}\right)^2 = \frac{b^2}{4a^2} - \frac{c}{a} = \frac{b^2}{4a^2} - \frac{4ac}{4a^2} = \frac{b^2-4ac}{4a^2}$$

taking square roots, we have

$$x+\frac{b}{2a} = \pm\sqrt{\frac{b^2-4ac}{4a^2}} = \frac{\pm\sqrt{b^2-4ac}}{2a}$$

Solving for x, we find

$$x = -\frac{b}{2a} \pm \frac{\sqrt{b^2-4ac}}{2a}$$

or writing as a single fraction we have

$$x = \frac{-b\pm\sqrt{b^2-4ac}}{2a} \qquad\qquad (2)$$

Quadratic Formula

The solution given in (2) is the solution to the general quadratic equation and is known as the *Quadratic formula*. We remind you that there are two solutions, as the $\pm$ symbol is a shorthand that tells us first use the $-$ sign to write one solution, and then use the $+$ sign to write the other solution. Sometimes, we call one solution x_1 and the second solution x_2. Sometimes, we call the solutions the *roots*.

We illustrate the use of this formula on the following exercises. First, some suggestions that will make the use of the formula more convenient.

Given any quadratic equation, perform the following steps, if required:

    1.   **Rewrite the equation so that the terms are in descending powers.**

    2.   **Rewrite the equation so that the coefficient of the squared term is positive.**

    3.   **Clear all fractions so the coefficients are all integers.**

We remark that Step 2 is easily accomplished by multiplying each term in the equation by -1, and Step 3 is accomplished by multiplying each term in the equation by the least common denominator. These remarks will be illustrated in the examples that follow.

Example 1
Use the quadratic formula to solve the equation $6x^2 + 5x = 6$.

Solution
We first rewrite the equation as $6x^2 + 5x - 6 = 0$. We identify the coefficients; $a = 6$, $b = 5$ and $c = -6$. We

next substitute into the quadratic formula to obtain

$$x = \frac{-5\pm\sqrt{(5)^2-4(6)(-6)}}{2(6)} = \frac{-5\pm\sqrt{25+144}}{12} = \frac{-5\pm\sqrt{169}}{12} = \frac{-5\pm13}{12}$$

therefore, we have as our two solutions (roots),

$$x_1 = (-5+13)/12 = 8/12 = 2/3 \text{ and } x_2 = (-5-13)/12 = -18/12 = -3/2.$$

■■■

Example 2
Use the quadratic formula to solve the equation $4x = -3x^2 + 8$.

Solution
We first apply Step 1 and rewrite the equation as $3x^2 + 4x - 8 = 0$. We now identify $a = 3$, $b = 4$, and $c = -8$. Substituting into the quadratic equation, we have

$$x = \frac{-4\pm\sqrt{(4)^2-4(3)(-8)}}{2(3)} = \frac{-4\pm\sqrt{16+96}}{6} = \frac{-4\pm\sqrt{112}}{6}$$

We next simplify this expression, remembering what we learned about the simplification of radicals. We have,

$$x = \frac{-4\pm\sqrt{16\cdot7}}{6} = \frac{-4\pm4\sqrt{7}}{6} = \frac{2(-2\pm2\sqrt{7})}{6} = \frac{-2\pm2\sqrt{7}}{3}$$

thus, we have irrational conjugate expressions as the two roots (solutions)

$$x_1 = \frac{-2+2\sqrt{7}}{3} \text{ and } x_2 = \frac{-2-2\sqrt{7}}{3}$$

■■■

If we needed numerical solutions, we would compute them using a calculator to approximate the square root.

Example 3
(a) Solve the quadratic equation $\frac{2}{3}x^2 - \frac{3}{4}x + \frac{1}{6} = 0$. (b) Approximate the roots to the nearest thousandth

Solution
(a) We first clear fractions by multiplying each term by the least common multiple 12. We obtain $8x^2 - 9x + 2 = 0$. We set $a = 8$, $b = -9$, and $c = 2$. This gives

$$x = \frac{-(-9)\pm\sqrt{(-9)^2-4(8)(2)}}{2(8)} = \frac{9\pm\sqrt{81-64}}{16} = \frac{9\pm\sqrt{17}}{16}$$

Thus, the two roots are

$$x_1 = \frac{9-\sqrt{17}}{16} \text{ and } x_2 = \frac{9+\sqrt{17}}{16}$$

(b) Using $\sqrt{17} \approx 4.1231056$ we obtain, to three decimal places, $x_1 = 0.305$ or $x_2 = 0.820$.

■■■

Clearing Fractions	**Example 4** Solve the quadratic equation $\frac{1}{4}z^2 + \frac{5}{8} = \frac{1}{2}z$.

Solution

We clear fractions by multiplying each term by the *LCD* which is 8, giving $2z^2 + 5 = 4z$. We rewrite this equation as $2z^2 - 4z + 5 = 0$. We set $a = 2$, $b = -4$ and $c = 5$. Substitution into the quadratic formula, gives

$$z = \frac{-(-4) \pm \sqrt{(-4)^2 - 4(2)(5)}}{2(2)} = \frac{4 \pm \sqrt{16 - 40}}{4} = \frac{4 \pm \sqrt{-24}}{4} =$$

$$\frac{4 \pm i\sqrt{24}}{4} = \frac{4 \pm i\sqrt{4}\sqrt{6}}{4} = \frac{4 \pm 2\sqrt{6}i}{4} = \frac{2(2 \pm \sqrt{6}i)}{4}$$

$$z = \frac{2 \pm \sqrt{6}i}{2}$$

Thus, there are two complex conjugate roots, $z_1 = \dfrac{2 - \sqrt{6}i}{2}$ and $z_2 = \dfrac{2 + \sqrt{6}i}{2}$.

■ ■ ■

Applications	In applications involving quadratic equations, we sometimes find that both solutions make sense and sometimes we find that one of the solutions needs to be rejected as it makes no physical sense. Consider the following examples.

Example 5

The altitude h of a rocket fired vertically upward is given by the equation $h = -16t^2 + 2500t$, where t is the time from firing in seconds. (a) How long does it take the rocket to reach an altitude of 4700 feet? (b) How long does it take for the rocket to return to the ground?

Solution

(a) We are asked to find t when $h = 4700$ feet. Therefore we need to solve the equation

$$4700 = -16t^2 + 2500t$$

We rewrite this equation as

$$16t^2 - 2500t + 4700 = 0$$

Using the quadratic formula, we have

$$t = \frac{2500 \pm \sqrt{2500^2 - 4 \cdot 16 \cdot 4700}}{32}$$

$$t = \frac{2500 \pm \sqrt{5949200}}{32}$$

Using a calculator, we find the two solutions are $t_1 \approx 154.347$ seconds or $t_2 \approx 1.903$ seconds. Why are there two solutions? When the rocket is going upward, it reaches a height of 4700 feet in approximately 1.903

seconds. However, this rocket will achieve its maximum altitude and then begin to descend. At approximately 154.347 seconds after launch it will again be at this position.

(b) When the rocket returns to the ground its altitude $h = 0$, so we need to solve the equation

$$0 = -16t^2 + 2500t$$

This equation may be solved by factoring

$$0 = -t(16t - 2500)$$

yielding $t = 0$ or $t = 2500/16 = 156.25$ seconds. When $t = 0$, the rocket is being launched, so the time it takes for the rocket to return to the ground is 156.25 seconds.
■■■

Exercise 6
One leg of a right triangle is 6 inches and the hypothenuse is 14 inches, find the length of the other leg of the triangle to the nearest thousandth of an inch.

Solution
Let x represent the length of the unknown leg. Using Pythagoras' theorem, we have

$$x^2 + 6^2 = 14^2$$

isolating the unknown, we find that

$$x^2 = 160,$$

$$x = \pm\sqrt{160} \approx = \pm 12.649$$

Since x represents a length, it must be a positive number, therefore, we reject the negative solution. The other leg of the triangle is therefore approximately 12.649 inches.
■■■

Equations reducible to Quadratics	Sometimes, we need to perform some algebraic simplifications on an equation before we recognize it as a quadratic, as the next two examples illustrate.

Exercise 7
Solve for x: (note that $x \neq -8, 2$. Why?)

$$\frac{24}{x+8} + \frac{20}{x-2} = 12.$$

Solution
We multiply each term on both sides of the equation by the *LCD* which is $(x + 8)(x - 2)$, yielding

$$24(x - 2) + 20(x + 8) = 12(x + 8)(x - 2)$$

Multiplying out and collecting similar terms, we obtain

$$12x^2 + 28x - 304 = 0$$

or

$$4(3x^2 + 7x - 76) = 4(x - 4)(3x + 19) = 0$$

Therefore $x_1 = -19/3$ and $x_2 = 4$. We leave it as an exercise for you to check the solutions.
∎∎∎

Sometimes, a trinomial may be at first glance appear not to be a quadratic equation, but by a simple substitution it may be transformed into a trinomial which is a quadratic, as illustrated by the next example.

Exercise 8

Solve the equation $8x^{\frac{3}{4}} - 19x^{\frac{3}{8}} - 27 = 0$.

Solution
The key to this problem and problems similar to it is that the highest power in the trinomial is the square of the next highest power. Thus, if we let the smallest power be called u, i.e.,

$$u = x^{\frac{3}{8}}$$

then

$$u^2 = \left(x^{\frac{3}{8}}\right)^2 = x^{\frac{3}{4}}$$

and we may rewrite the original equation as

$$8u^2 - 19u - 27 = 0$$

which is a quadratic in the variable u and factors as

$$(8u - 27)(u + 1) = 0$$

Thus $u = 27/8$ or $u = -1$.

However, we need to find x. Since $u = x^{\frac{3}{8}}$, we have $(u)^{\frac{8}{3}} = (x^{\frac{3}{8}})^{\frac{8}{3}}$ or $x = u^{\frac{8}{3}}$. Therefore, when $u = -1$, we have $x = (-1)^{8/3} = 1$; when $u = 27/8$, we have $x = (27/8)^{8/3} = ((27/8)^{1/3})^8 = (3/2)^8 = 6561/256$. Thus, we have $x = 1$ or $x = 6561/256$.
∎∎∎

Exercise set 11

In exercises 1-15 rewrite the equation in the form $ax^2 + bx + c = 0$, with $a > 0$, and then determine the coefficients a, b, and c. Do not solve the equation.
1. $x^2 + 4x - 11 = 0$ 3. $5x^2 - 8x - 12 = 0$

2. $3x^2 + 7x - 8 = 0$ 4. $7x + 3x^2 - 10 = 0$

5. $-27x + 10 - 5x^2 = 0$

6. $4x = 9x^2 + 4$

7. $3 - 2x = 8x + 5x^2 - 11$

8. $2x^2 = 10$

9. $5x^2 = -11$

10. $3x = 7x^2$

11. $4x^2 = 9x$

12. $(x - 2)(3x + 4) = 5$

13. $(2x - 5)(4x + 3) = 7$

14. $4(3x - 4)(5x - 6) = 25$

15. $(x - 7)(2x - 3) = 3x(2x - 4)$

In each of the following, solve the given quadratic equation exactly using the quadratic formula. Write the solutions in its simplest form. Using a calculator, determine all irrational solutions to the nearest thousandth.

16. $x^2 + 2x - 10 = 0$

17. $x^2 + 6x + 3 = 0$

18. $x^2 - 2x - 10 = 0$

19. $x^2 - 6x + 3 = 0$

20. $x^2 + 6x = 3$

21. $x^2 - 8x = 4$

22. $x^2 - 12x - 4 = 0$

23. $x^2 + 5x = 5$

24. $x^2 - 5x = 5$

25. $x^2 - 3x - 1 = 0$

26. $x^2 + 2x + 10 = 0$

27. $(x - 4)(x + 1) = 6$

28. $(x - 3)(x + 5) = 20$

29. $(2x - 3)(3x + 1) = -4$

30. $(x - 3)(x + 5) = 12$

31. $(x + 3)(x + 2) = -10$

32. $(2x - 3)(3x + 5) = -20$

33. $x^2 - 6x + 23 = 0$

34. $x^2 - 12x = -48$

35. $x^2 + 10x + 50 = 0$

36. $x^2 - 8x + 36 = 0$

37. $x^2 - 4x = 8$

38. $x^2 + 6x + 21 = 0$

39. $x^2 - 3x + 3 = 0$

40. $x^2 + 5x + 10 = 0$

41. $2x^2 + 8x + 1 = 0$

42. $2x^2 + x - 6 = 0$

43. $6x^2 + 5x - 6 = 0$

44. $12x^2 - 7x = 12$

45. $6x^2 + 23x = -20$

46. $2x^2 + 5x + 10 = 0$

47. $4x^2 - 5x + 8 = 0$

48. $\frac{1}{2}x^2 + 5x + 2 = 0$

49. $\frac{1}{2}x^2 - \frac{3}{4}x = \frac{5}{6}$

50. $\frac{3}{5}x^2 - \frac{3}{2}x = \frac{7}{10}$

51. $3 - 4x^2 = 8x$

52. $9 - 3x^2 + 5x = 8 + 2x^2$

53. The sum of the squares of three consecutive integers is 110, find the integers.

54. The sum of the squares of three consecutive odd integers is 515, find the integers.

55. If the legs of an isosceles right triangle are each 12 inches, how long, to the nearest one-thousandth of an inch, is the hypothenuse?

56. The hypothenuse and one leg of a right triangle are 18 and 12 inches respectively, how long, to the nearest one-thousandth of an inch is the other leg?

57. A ball is thrown vertically upward from the ledge of a building 75 feet above ground. The ball's height h in feet above the ground at time t in seconds is given by the equation

$$h = -16t^2 + 80t + 75.$$

(a) how long does it take the ball to reach a height of 90 feet? (b) Howlong before the ball is back to its original position (at 75 feet)? (c) How long before the ball hits the ground? Give each answer to the nearest one-thousandth of an inch.

58. An object is dropped from a helicopter hovering at 250 feet above the ground. The objects height in feet, h is given by the equation $h = -16t^2 + 250$, where t is measured in seconds. How long before the object is (a) 100 feet above the ground? (b) 50 feet above the ground? (c) on the ground? Give each answer to the nearest one-thousandth of an inch.

59. A rectangular swimming pool is 30 feet by 40 feet. If a rectangular strip of grass of uniform width is to go around the pool, and the area of this strip is 624 square feet, how wide is the strip?

60. Two cars leave an intersection at the same time, one goes north and the other goes east. Some time later they are 125 miles apart. If the car moving north traveled 12 miles more than the one going east, how many far (to the nearest mile) did each car travel?

61. Barbara wants to purchase an area rug for her dining room whose dimensions are 20 feet by 24 feet. If the rectangular rug she purchases has an area of 216 square feet and is placed an equal distance from each wall (a)how wide is the uniform strip of uncovered flooring? (b) what are the dimensions of the rug?

62. John bikes a distance of 120 miles and then returns over the same route. On his return his average speed is 2 miles per hour more than when going. If the combined time for both trips was 22 hours, what was his speed each way?

63. Mary can build a computer in two hours less time than Tim. Working together, they can build a computer in 2 hours and 24 minutes. How long does it take Mary alone to build a computer?

Solve each of the following equations for *the real values of x.*

64. $\dfrac{12}{x-4} + \dfrac{18}{x+4} = \dfrac{9}{2}$

65. $\dfrac{18}{2x+3} - \dfrac{8}{x+5} = 1$

66. $\dfrac{4x}{3x-2} + \dfrac{16-3x}{2x+2} = 2$

67. $x^4 + 5x^2 - 36 = 0$

68. $6x^6 - 17x^3 + 12 = 0$

69. $x^{\frac{2}{3}} - 35x^{\frac{1}{3}} + 216 = 0$

70. $2x^4 - 4x^2 - 3 = 0$

71. Here is another proof of the quadratic formula. Begin with $ax^2 + bx + c = 0$, and multiply each term of the equation by $4a$. This gives $4a^2x^2 + 4abx + 4ac = 0$. Rewrite the equation as $4a^2x^2 + 4abx = -4ac$ and add b^2 to each side giving $4a^2x^2 + 4abx + b^2 = b^2 - 4ac$. Now factor the left-hand-side of this equation and complete the proof.

72. Given the quadratic equation $ax^2 + bx + c = 0$ and $cx^2 + bx + a = 0$, prove that the roots of one equation are the reciprocals of the roots of the other equation.

POSTTEST 11 - Time 10 minutes

Solve each of the following using the quadratic formula. Leave answers in simplest radical form. Each question is worth two points.

1. $4x^2 - 3x - 2 = 0$ 2. $x - 2x^2 = -4$ 3. $\frac{2}{5}x^2 = \frac{1}{10}x + \frac{1}{2}$ 4. $4x^2 - 5x + 3 = 0$

5. A ball is thrown vertically upward from a ledge of a building 200 feet above the ground. Its height above the ground s, measured in feet, in terms of its time of flight t, measured in seconds, is given by the equation $s = -16t^2 + 86t + 200$. How long does it take the ball to hit the ground.

(Notes)

12. Observations and Extensions

» **Discriminant**
» **Sum and Product of Roots**

PRETEST 12-Time 10 minutes

Each question is worth two points.

Without solving, determine the nature of the solutions (roots) of the quadratic equations.

1. $6x^2 - 13x + 6 = 0$

2. $x^2 - 4x - 6 = 0$

3. Determine a quadratic equation whose highest order term has coefficient 1, if its two solutions are $x = 3$ and $x = -2$.

Given the quadratic equation $3x^2 - 2x + 5 = 0$, determine, without solving the

4. Sum of the roots

5. Product of the roots.

In the last section, we showed that the solutions (roots) to the quadratic equation $ax^2 + bx + c = 0$ are given by the quadratic formula

$$x = \frac{-b \pm \sqrt{b^2 - 4ac}}{2a} \tag{1}$$

Discriminant

The term appearing in the square root, describes the nature of the solutions and is called the discriminant, denoted by D. That is, we define

$$D = b^2 - 4ac \tag{2}$$

then we may rewrite the quadratic formula as

$$x = \frac{-b \pm \sqrt{D}}{2a} \tag{3}$$

There are three cases to consider: (I) $D = 0$, (II) $D > 0$ and (III) $D < 0$.

(I) If $D = 0$, then (1) yields one solution to the quadratic equation, namely $x = -\frac{b}{2a}$. Such a solution is called a *double* or *repeated* root. Consider the quadratic equation

$$4x^2 + 12x + 9 = 0$$

Its discriminant, $D = (12)^2 - 4(4)(9) = 0$. That means its solution is $x = -12/((2)(4)) = -3/2$. Of course, if we factored the equation we have

$$(2x + 3)(2x + 3) = (2x + 3)^2 = 0$$

We see immediately that $x = -3/2$ is the double root (repeated root) because there are two identical factors leading to the same solution.

(II) If $D > 0$ then we see from (1) that there will be two solutions to the quadratic equation. This case can be examined a little more closely if we divide it into two sub-cases, namely when D is a perfect square and when it is not.

When D is a perfect square, then its square root is a rational number and it then follows that both solutions are rational numbers. For example, consider the equation $6x^2 + 5x - 6 = 0$. Its discriminant, $D = (5)^2 - 4(6)(-6) = 169$, a perfect square and from (3), we have that

$$x = \frac{-5 \pm \sqrt{169}}{2 \cdot 6} = \frac{-5 \pm 13}{12}$$

yielding as its two rational roots, $x = -3/2$ and $x = 2/3$.

When D is not a perfect square, then we obtain two conjugate irrational roots as solutions to the quadratic equation. For example, consider the equation $2x^2 - 4x - 3 = 0$, its discriminant is

$$D = (-4)^2 - 4(2)(-3) = 40$$

This is not a perfect square and the solutions to the quadratic equation are

$$x = \frac{4 \pm \sqrt{40}}{4} = \frac{2 \pm \sqrt{10}}{2}$$

Note the two roots are $x_1 = \frac{2 - \sqrt{10}}{2}$ and $x_2 = \frac{2 + \sqrt{10}}{2}$, conjugate irrational roots.

(III) When the discriminant is negative, then the radicand is negative and we have two complex conjugate roots as the solutions to the quadratic equation. As an illustration, consider

$$2x^2 - 4x + 3 = 0$$

its discriminant is

$$D = (-4)^2 - 4(2)(3) = -8$$

and the solutions are

$$x = \frac{4 \pm \sqrt{-8}}{4} = \frac{4 \pm 2\sqrt{2}i}{4} = \frac{2 \pm \sqrt{2}i}{2}$$

Note the two roots are $x_1 = \dfrac{2 - \sqrt{2}\,i}{2}$ and $x_2 = \dfrac{2 + \sqrt{2}\,i}{2}$, conjugate complex roots.

We summarize, the three cases in Table 1.

Discriminant $D = \sqrt{b^2 - 4ac}$	Nature of the Roots of the Quadratic
Negative	Two complex conjugate roots
Zero	One double (repeated) root
Positive and a perfect square	Two rational roots
Positive and not a perfect square	Two irrational conjugate roots

Table 1 - Nature of the Roots of a Quadratic Equation

(Note that when D is zero or a perfect square then usually the fastest way to solve the quadratic equation is by factoring.)

Example 1
Discuss the nature of the roots of the quadratic equation $2x^2 - 5x - 2 = 0$.

Solution
The discriminant $D = (-5)^2 - 4(2)(-2) = 41$, a real number which is not a perfect square, therefore the roots are irrational conjugates.
■■■

Example 2
Discuss the nature of the roots of the quadratic equation $2x^2 - 5x + 6 = 0$.

Solution
The discriminant $D = (-5)^2 - 4(2)(6) = -23$, therefore the roots are complex conjugates.
■■■

Example 3
Discuss the nature of the roots of the quadratic equation $4x^2 - 20x + 25 = 0$.

Solution
The discriminant $D = (-20)^2 - 4(4)(25) = 0$, therefore there is a double (repeated) root.
■■■

Example 4
Discuss the nature of the roots of the quadratic equation $2x^2 - 3x - 5 = 0$

Solution
The discriminant $D = (-3)^2 - 4(2)(-5) = 49$, a perfect square, therefore the roots are both rational numbers.
■■■

Suppose you are given the roots of a quadratic, for example, suppose you are told that the two roots are $x_1 = 2$ and $x_2 = 3$, then working backwards, we have the quadratic in factored form as

$$(x - 2)(x - 3) = 0$$

Similarly, if the roots of a quadratic are $x_1 = 2/3$ and $x_2 = -3/2$, then working backwards, we have the quadratic equation

$$(x - 2/3)(x + 3/2) = 0$$

Sum and Product of Roots

More generally, given the quadratic equation $ax^2 + bx + c = 0$, divide by a (so the leading coefficient is 1) to obtain

$$x^2 + \frac{b}{a}x + \frac{c}{a} = 0 \tag{4}$$

Let us suppose we have solved this equation and denote the two solutions (roots) by x_1 and x_2 then we may rewrite (4) in factored form as

$$(x - x_1)(x - x_2) = 0$$

If we multiply this out and collect terms, we have

$$x^2 - (x_1 + x_2)x + x_1 x_2 = 0 \tag{5}$$

Comparing (4) and (5), we have for the quadratic equation $ax^2 + bx + c = 0$, whose roots are x_1 and x_2,

$$x_1 + x_2 = -\frac{b}{a}$$

$$x_1 x_2 = \frac{c}{a} \tag{6}$$

In words, this says something very interesting about the relationship between the roots and coefficients of a quadratic equation. Namely, the sum of the roots is $-\frac{b}{a}$ and their product is $\frac{c}{a}$.

Example 5
Given the quadratic equation $2x^2 - 3x - 5 = 0$, determine (a) the sum of its roots and (b) the product of its roots.

Solution
(a) $a = 2$, $b = -3$ and $c = -5$, therefore the sum of the roots is $-b/a = -(-3)/2 = 3/2$. (b) the product of the roots is $c/a = (-5)/2 = -5/2$.
■■■

Example 6

Find a quadratic equation with integer coefficients if the sum of its roots is 5 and their product is 2/3.

Solution

We have that $-b/a = 5$, therefore $b/a = -5$, and $c/a = 2/3$. Substituting into (4), we have

$$x^2 - 5x + \frac{2}{3} = 0$$

since we want all the coefficients to be integers, we multiply by the *LCD* which is 3, to obtain

$$3x^2 - 15x + 2 = 0$$

(alternately, we may substitute directly into (5) to obtain the equation.)

∎∎∎

We remark that rather than checking both roots of a quadratic equation it is often more convenient to use (6).

Exercise set 12

In Exercises 1- 25, (a) examine the discriminant, determine the nature of the roots of the given quadratic equation, (b) very your result by solving the quadratic equation.

1. $x^2 - 18x + 81 = 0$

2. $3x^2 - 6x + 3 = 0$

3. $2x^2 - 16x + 32 = 0$

4. $-4x^2 + 24x - 36 = 0$

5. $5x^2 + 20x = -20$

6. $\frac{1}{5}x^2 + 2x + 5 = 0$

7. $\frac{1}{9}x^2 + \frac{2}{3}x + 1 = 0$

8. $6x^2 = x + 12$

9. $10x^2 + 13x = 3$

10. $12x^2 = 12 - 7x$

11. $14x^2 + 29x = 15$

12. $2x^2 + 7x = -3$

13. $-16x + 35 = 12x^2$

14. $x^2 - \frac{13}{3}x + \frac{10}{3} = 0$

15. $3x^2 - 8x + 2 = 0$

16. $2x^2 + 5x + 1 = 0$

17. $3x^2 + 2x - 7 = 0$

18. $5x^2 + 8x - 2 = 0$

19. $x^2 = 5x + 5$

20. $\frac{1}{3}x^2 - \frac{1}{4}x - \frac{2}{5} = 0$

21. $x^2 - 2x + 3 = 0$

22. $2x^2 - 3x - 2 = 0$

23. $3x^2 + 4x + 2 = 0$

24. $-5x^2 + 4x - 2 = 0$

25. $8x^2 + 5x + 3 = 0$

In Exercises 26 -46, find a quadratic equation with integer coefficients whose roots are x_1 and x_2 as given.

26. $x_1 = 3, x_2 = 5.$

27. $x_1 = 4, x_2 = 7.$

28. $x_1 = -2, x_2 = 5.$

29. $x_1 = 6, x_2 = -7.$

30. $x_1 = 3, x_2 = 3.$

31. $x_1 = 5, x_2 = 5.$

32. $x_1 = -3, x_2 = -3.$

33. $x_1 = 3/2, x_2 = 5/3.$

34. $x_1 = -2/3, x_2 = -1/4$

35. $x_1 = 1/5, x_2 = -3/7.$

36. $x_1 = 1 - \sqrt{2}, x_2 = 1 + \sqrt{2}.$

37. $x_1 = 2 - 3\sqrt{5}, x_2 = 2 + 3\sqrt{5}.$

38. $x_1 = 5 - 2\sqrt{6}, x_2 = 5 + 2\sqrt{6}.$

39. $x_1 = 3 - 4\sqrt{5}, x_2 = 3 + 4\sqrt{5}.$

40. $x_1 = 3 + i, x_2 = 3 - i.$

41. $x_1 = 1 - 2i, x_2 = 1 + 2i.$

42. $x_1 = 5 - 3i, x_2 = 5 + 3i.$

43. $x_1 = 1 - \sqrt{2}i, x_2 = 1 + \sqrt{2}i.$

44. $x_1 = 2 - 3\sqrt{5}i, x_2 = 2 + 3\sqrt{5}i.$

45. $x_1 = 5 - 2\sqrt{6}i, x_2 = 5 + 2\sqrt{6}i.$

46. $x_1 = 3 - 4\sqrt{5}i, x_2 = 3 + 4\sqrt{5}i.$

In Exercises 47 - 71, find (a) the sum and (b) the product of the roots for the given quadratic equation.

47. $x^2 - 18x + 81 = 0$

48. $3x^2 - 6x + 3 = 0$

49. $2x^2 - 16x + 32 = 0$

50. $4x^2 - 24x + 36 = 0$

51. $5x^2 + 20x + 20 = 0$

52. $\frac{1}{5}x^2 + 2x + 5 = 0$

53. $\frac{1}{9}x^2 + \frac{2}{3}x + 1 = 0$

54. $6x^2 - x - 12 = 0$

55. $10x^2 + 13x - 3 = 0$

56. $12x^2 + 7x - 12 = 0$

57. $14x^2 + 29x - 15 = 0$

58. $2x^2 + 7x + 3 = 0$

59. $12x^2 + 16x - 35 = 0$

60. $x^2 - \frac{13}{3}x + \frac{10}{3} = 0$

61. $3x^2 - 8x + 2 = 0$

62. $2x^2 + 5x + 1 = 0$

63. $3x^2 + 2x - 7 = 0$

64. $5x^2 + 8x - 2 = 0$

65. $x^2 - 5x - 5 = 0$

66. $\frac{1}{3}x^2 - \frac{1}{4}x - \frac{2}{5} = 0$

67. $x^2 - 2x + 3 = 0$

68. $2x^2 - 3x - 2 = 0$

69. $3x^2 + 4x + 2 = 0$

70. $5x^2 - 4x + 2 = 0$

71. $8x^2 + 5x + 3 = 0$

72.   Prove the results given in (5) directly from the quadratic formula.

73.   (a) Given the cubic equation
        $$ax^3 + bx^2 + cx + d = 0,$$
        Suppose it has three roots designated by $x_1, x_2,$ and x_3. Find relationships between the coefficients of the cubic equation and its three roots. Hint: rewrite the equation as
        $$a(x - x_1)(x - x_2)(x - x_3) = 0.$$

        (b) Generalize your results to the nth order polynomial

        $$a_n x^n + a_{n-1} x^{n-1} + a_{n-2} x^{n-2} + \cdots + a_1 x + a_0 = 0$$

74.   Suppose A and B are real numbers and $C = c_1 + c_2 i$, where c_1 and c_2 are real numbers. Under what conditions will the roots of the quadratic equation $Ax^2 + Bx + C = 0$ have real solutions.

75.   Can any simple conclusions be made about the nature of the roots of the quadratic equation $Ax^2 + Bx + C = 0$ Where A, B and C are *complex* numbers?

POSTTEST 12-Time 10 minutes

Each question is worth two points.

Without solving, determine the nature of the solutions (roots) of the quadratic equations.

1. $2x^2 - 5x + 4 = 0$ 2. $3x^2 - 2x - 5 = 0$

3. Determine a quadratic equation whose highest order term has coefficient 1, if its two solutions are $x = 5$ and $x = -6$.

Given the quadratic equation $7x^2 - 5x - 3 = 0$, determine, without solving the

4. Sum of the roots   5. Product of the roots.

120

(Notes)

13. Equations Containing Radicals

» **Isolation and Extraneous Roots**
» **Multiple Square Roots**

PRETEST 13- Time 10 minutes

Each question is worth two points.

Solve for x:

1. $\sqrt{x} - 3 = 1$
2. $\sqrt{3x + 10} - 2 = 2$
3. $\sqrt{3x + 1} = 2x - 6$

4. $\sqrt{2x^2 + 7} = 5$
5. $\sqrt{3x + 1} - \sqrt{3x - 2} = 1$

The basic problem considered in this section is the solution of equations involving one or more square roots. Actually, equations involving other roots are often solved in a similar way, as we saw in Section 6, but in this section we shall examine only square roots.

The procedure is very simple. Given any equation involving square roots, isolate a square root term on one side of the equation and then square each side of the equation. Squaring a square root eliminates the square root. However, we have to be careful, sometimes the resulting equation will have solutions that do not satisfy the original equation, therefore we must always check the candidates for the solution in original equation. Those solutions to the modified equation that do not satisfy the original equation are sometimes called *extraneous roots*. These extraneous roots arise when the original equation is squared. All solutions to the original equation are solutions to the modified equation, but the reverse may not be true. Consider the equation

Isolation and Extraneous Roots

$$x = 3$$

squaring both sides gives

$$x^2 = 9$$

The solutions to this equation are $x = -3$ and $x = 3$. Only the second of these solutions, $x = 3$ satisfies the original equation, the first, $x = -3$ does not.

We illustrate the procedure with a simple illustration consider the equation

$$\sqrt{x} - 2 = 0$$

If we isolate the square root, we have

$$\sqrt{x} = 2$$

if we square both sides and solve, we have

$$\sqrt{x} = 2$$
$$\left(\sqrt{x}\right)^2 = 2^2$$
$$x = 4$$

Of course, we must check that $x = 4$ does satisfy the original equation. We have

$$\sqrt{4} - 2 = 2 - 2 = 0$$

so $x = 4$ is indeed the solution to the original equation.

Sometimes it is clear, without doing any work that a problem has no real solution. For example consider the equation $\sqrt{x} = -4$, this equation has no real solution since the square root of any real number is positive.

The are minor variations on the procedure which we illustrate in the examples that follow.

Example 1
Solve the equation $\sqrt{2x + 3} - 3 = 0$.

Solution
We first isolate the radical and we have

$$\sqrt{2x + 3} = 3$$

We next square both sides and then solve the resulting equation for x.

$$(\sqrt{2x + 3})^2 = 3^2$$
$$2x + 3 = 9$$
$$2x = 6$$
$$x = 3$$

We next check to make certain that $x = 3$ is the solution to the original equation.

$$\sqrt{2(3) + 3} = \sqrt{9} = 3$$

It does, therefore the solution is indeed $x = 3$.
∎∎∎

Example 2

Solve the equation $\sqrt{5x+1} = 4x-8$.

Solution

The square root term is already isolated, so we need only square both sides and solve.

$$(\sqrt{5x+1})^2 = (4x-8)^2$$
$$5x+1 = 16x^2-64x+64$$
$$16x^2-69x+63 = 0$$
$$(16x-21)(x-3) = 0$$
$$x = \frac{21}{16} \text{ or } x = 3$$

We now must check the two possibilities, first we check $x = 21/16$.

$$\sqrt{5\left(\frac{21}{16}\right)+1} \overset{?}{=} 4(\frac{21}{16})-8$$
$$\sqrt{\frac{121}{16}} \overset{?}{=} (\frac{21}{4})-8$$
$$\frac{11}{4} \neq -\frac{11}{4}$$

Clearly this does not check, therefore 21/16 is not a solution to the original equation. We next check to see if $x = 3$ checks the original equation.

$$\sqrt{5(3)+1} \overset{?}{=} 4(3)-8$$
$$4 = 4$$

This does check, therefore we have as the solution to this problem, $x = 3$.

■■■

Example 3

Solve the equation $\sqrt{3x^2+4} = 4$.

Solution

$$\sqrt{3x^2+4} = 4$$
$$(\sqrt{3x^2+4})^2 = 4^2$$
$$3x^2+4 = 16$$
$$3x^2 = 12$$
$$x^2 = 4$$
$$x = \pm\sqrt{4} = \pm2$$

Thus, we have two possible solutions $x = -2$ and $x = 2$. We leave it to you to verify that they both check the original equation, so our two solutions are indeed $x = -2$ and $x = 2$.

■■■

Multiple square roots

If there is more than one square root in the problem, then we isolate one of them, and square the equation. However, when this is done, there may still be a square root present. What do we do? We repeat the procedure, isolate the remaining square root and square again. We illustrate this in the next example.

Example 4

Solve the equation $\sqrt{4x+1} - \sqrt{2x-8} = 3$.

Solution

We isolate one of the square roots.

$$\sqrt{4x+1} = 3 + \sqrt{2x-8}$$
$$(\sqrt{4x+1})^2 = (3+\sqrt{2x-8})^2$$
$$4x+1 = 9 + 6\sqrt{2x-8} + (2x-8)$$
$$2x = 6\sqrt{2x-8}$$

We combined like terms and isolated the square root in this last step. We once again square both sides.

$$(2x)^2 = (6\sqrt{2x-8})^2$$
$$4x^2 = 36(2x-8)$$
$$4x^2 - 72x + 288 = 0$$
$$4(x^2 - 18x + 72) = 0$$
$$(x-12)(x-6) = 0$$

Thus, there are two possibilities, namely $x = 6$ or $x = 12$. We next must determine if they are both solutions to the original equation.

$$\sqrt{4(6)+1} - \sqrt{2(6)-8} \overset{?}{=} 3$$
$$\sqrt{25} - \sqrt{4} \overset{?}{=} 3$$
$$5 - 2 = 3$$

This checks, so $x = 6$ is a solution.

$$\sqrt{4(12)+1} - \sqrt{2(12)-8} \overset{?}{=} 3$$
$$\sqrt{49} - \sqrt{16} \overset{?}{=} 3$$
$$7 - 4 = 3$$

$x = 12$ also checks. Therefore, both $x = 6$ and $x = 12$ are solutions to this problem.

■■■

Example 5

If four times the square root of the sum of the number and 3 is twice the number, find the number.

Solution

We first translate the problem into an algebraic equation. We let x be the number. Then we have

$$4\sqrt{(x+3)} = 2x$$

We square both sides and obtain

$$16(x+3) = 4x^2$$
$$4x^2 - 16x - 48$$
$$4(x^2 - 4x - 12) = 0$$
$$(x-6)(x+2) = 0$$
$$x = 6 \text{ or } x = -2$$

x cannot be negative since the left hand side of the equation is always positive, so it is clear from this observation that $x = -2$ will not check. We leave it for you to verify that $x = 6$ does check and therefore it is the solution to the problem.

■■■

Exercise set 13

Solve for the real values of the unknown in Exercises 1 - 25. Check your answers!

1. $\sqrt{x} = 5$

2. $\sqrt{x} = 3$

3. $\sqrt{x+1} = 6$

4. $\sqrt{2x} = 4$

5. $\sqrt{5x} = 7$

6. $\sqrt{3x+13} = 5$

7. $\sqrt{5x+6} = 6$

8. $\sqrt{7x-5} = 3$

9. $\sqrt{2x+3} = 5x - 12$

10. $\sqrt{7x+8} + 2x = 5x - 16$

11. $\sqrt{8x-7} + 2x = 6x - 11$

12. $\sqrt{6-5x} - 2x = 3x + 14$

13. $\sqrt{3x-5} + 2x + 3 = 5x - 2$

14. $x = \sqrt{x^2 - 4x + 20}$

15. $x = \sqrt{x^2 - 10x + 40}$

16. $x = \sqrt{x^2 + 5x - 30}$

17. $x = \sqrt{x^2 + 3x - 18}$

18. $\sqrt{2x^2 + 7} = 3x - 4$

19. $\sqrt{4x^2 + 5} = 12x - 9$

20. $\sqrt{4 - 3x^2} = 11x - 10$

21. $\sqrt{2x^2 + 2x + 1} = 2x - 1$

22. $\sqrt{x+5} + \sqrt{x} = 5$

23. $\sqrt{4x+5} = \sqrt{x+4} + 2$

24. $3\sqrt{2x+3} - \sqrt{x-2} = 8$

25. $\sqrt{2x^2 + 7} - \sqrt{9x^2 + 7} = -1$

26.   What is wrong with the following solution?

$$\sqrt{2x + 3} + \sqrt{3x - 2} = 6$$
$$2x + 3 + 3x - 2 = 36$$
$$5x + 1 = 36$$
$$5x = 35$$
$$x = 7$$

27.   Without solving explain why the following equation has no real solution.

$$\sqrt{3 - 2x} = -3$$

What happens if you solve the equation for x?

28.   The square root of some number is twice the number. Find the number.

29.   The square root of the sum of a number and 8 is one more than the square root of 5 less than the number. Find the number.

30.   The sum of a number and is square root is 5 more than the number. Find the number

31.   If twice the square root of some number is subtracted from three times the number, the result is 24 more than twice the number. Find the number.

32.   The distance between the points $(x, 6)$ and $(3, 2)$ is one-half the distance between $(x, 6)$ and $(12,-2)$. Find x.

33.   The current through a resistor is related to the power dissipated by the formula

$$I = \sqrt{\frac{P}{R}}$$

where the current I is measured in amperes, the power P in watts and the resistance R in ohms. Find the power dissipated through a 100 ohm resistor with a 5 ampere current.

34.   The minimum velocity that a rocket must have in order to escape the Earth's gravitational attraction is called the *escape* velocity. In can be shown that this velocity, v_e is given by the formula

$$v_e = \sqrt{2gR,}$$

Where $g = 32$ ft/sec² is the acceleration due to gravity and $R = 4000$ miles, is the approximate radius of the Earth. (a) Show that if we change the units of distance from feet to miles, (5280 feet = 1 mile), g= 0.00606061 miles/sec². (b) Find v_e in miles per second.

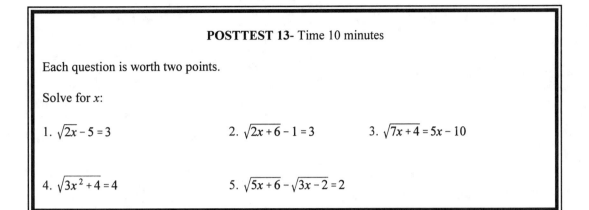

POSTTEST 13- Time 10 minutes

Each question is worth two points.

Solve for x:

1. $\sqrt{2x} - 5 = 3$

2. $\sqrt{2x + 6} - 1 = 3$

3. $\sqrt{7x + 4} = 5x - 10$

4. $\sqrt{3x^2 + 4} = 4$

5. $\sqrt{5x + 6} - \sqrt{3x - 2} = 2$

(Notes)

14. Solving Non- Linear Inequalities

» **Sign Analysis**
» **Interval Notation**

PRETEST 14- Time 15 minutes

Each question is worth two points. Solve the given inequality

Simplify:

1. $3x^2 + 10x \geq 8$ 2. $2x^3 - 12x + 18x < 0$ 3. $\dfrac{3 - 2x}{x - 5} \leq 0$

4. $\dfrac{2x - 3}{3x + 4} > 2$ 5. $\dfrac{2x^3(x^2 - 3x + 2)}{x^2 - 16} \leq 0$

Given the linear inequality, $3x - 2 > 0$, we wish to find all values of x for which this inequality is a true statement. We recall that the rules for manipulating an inequality are the same as those for solving an equation with one exception, namely when an inequality is multiplied or divided by a negative number the inequality symbol reverses. Thus, to solve the inequality $3x - 2 > 0$, we could write $3x > 2$ or $x > 2/3$. This inequality is completely solved, that is, for any x-value greater than 2/3, the inequality is true. Let us now re-examine the same problem and suggest another approach which will work on any inequality which may be factored and is to be less than or greater than zero.

We first solve $3x - 2 = 0$. This is of course yields $x = 2/3$. We draw the number line and indicate this value on the line as indicated in Figure 1.

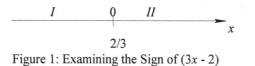

Figure 1: Examining the Sign of $(3x - 2)$

129

The only time the term $3x - 2$ is zero is when $x = 2/3$. For any other value of x it is either positive or negative. Therefore in each of the Regions I or II the sign must always be the same. Therefore, we need only choose *any* test value for x in each of these regions and examine the sign of $3x - 2$ at the test value. Consider Table 1 where we examine the sign at a test value in each region .

Region	Test Value	sign of $(3x - 2)$
I	0	$3(0) - 2 = -$
II	1	$3(1) - 2 = +$

Table 1: Sign of $(3x - 2)$

Thus, we see that in region II the expression $(3x - 2)$ is positive therefore our solution is $x > 2/3$.

We now generalize the second method and consider the following non-linear inequality in factored form

$$(x - 5)(x + 2) < 0$$

Our problem is to find those value of x which satisfy the inequality, that is, find those values of x for which the product of the two factors is negative. We first observe that the right hand side of the inequality is the number zero. This is important as we shall see.

Sign Analysis

Observe that the only time the product on the left-hand-side is zero is when either factor is zero, that is, when $x = -2$ or $x = 5$. That means for any other value of x the product of these factors must always be positive or negative. A simple way to determine this is as follows: draw the number line and indicate these two values for x on the number line as indicated in Figure 2.

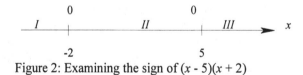

Figure 2: Examining the sign of $(x - 5)(x + 2)$

Note that at $x = -2$ and $x = 5$, we put a zero in the diagram. That is to indicate that the product of the two factors is zero at each of those x values. Moreover, we included letters for the three regions they created. Region I corresponds to the region in which $x < -2$, Region II corresponds to $-2 < x < 5$ and Region III to $x > 5$. Since the product is only zero at the indicated two x-values, it will be of the same sign, that is, only positive or only negative in the three indicated regions. All we need to do is to select *any* x-value in each of these regions and determine the sign of the product. We do this in Table 2.

Region	Test value	Sign of Product $(x-5)(x+2)$
I	$x = -3$	$(-8)(-1) = +$
II	$x = 1$	$(-4)(3) = -$
III	$x = 6$	$(1)(8) = +$

Table 2: Sign of $(x-5)(x+2)$

We can now redraw the sign diagram in Figure 2, indicating the sign of the product in each of the three regions. This is done in Figure 3.

Figure 3: Sign Diagram for $(x - 5)(x + 2)$

Interval Notation

This tells us that the solution to the inequality $(x - 5)(x + 2) < 0$ is any x in the interval $-2 < x < 5$, because in that region the product is negative. Sometimes the answer is written as $(-2, 5)$, this is shorthand for the interval $-2 < x < 5$. More generally, (a, b) is shorthand for the interval $a < x < b$, $[a, b]$ is shorthand for the interval $a \leq x \leq b$, and $(a, b]$ is shorthand for $a < x \leq b$. Note that the bracket is used to include the endpoint and the parenthesis is used to exclude it. (How would you indicate $a \leq x < b$ in interval notation?)

Notice that the given inequality could have been written as $x^2 - 3x - 10 < 0$, or $x^2 - 3x < 10$, or some other equivalent formulation could have been given. All that we need do is rewrite the expression so that one side of the inequality is the number zero, and then proceed as above.

The above suggests a general procedure for the solution of an inequality.

Solving an Inequality

1. Rewrite the inequality so that the expression involving x is on one side of the inequality symbol and the number 0 is on the other side of the inequality symbol.

2. Factor the expression involving x. If it is a fraction, factor both the numerator and denominator.

3. Locate the values of x at which each factor in both the numerator and denominator is zero.

4. Draw a number line indicating those x-values (in increasing order) found in Step 3. Put the number 0 in the sign diagram above each of the zeros of the numerator and *ND* for each zero of the denominator.

5. Choose any x-value in each of the regions separated by the values found in Step 3. Compute the sign of the expression at each of these values and enter them on the sign diagram.

6. Read the solution from the sign diagram.

We illustrate the procedure on the following examples.

Example 1
Determine the values of x satisfying the inequality (a) $2x^3 - 8x^2 \le -8x$ (b) $2x^3 - 8x^2 \ge -8x$.

Solution
(a) First, we rewrite the inequality so that zero is on one side. We have

$$2x^3 - 8x^2 + 8x \le 0$$

We want to determine when the expression on the left will be negative or zero. Factor the left hand side, to obtain

$$2x(x^2 - 4x + 4x) \le 0$$

or

$$2x(x - 2)(x - 2) \le 0$$

or finally,

$$2x(x - 2)^2 \le 0$$

The left hand side will be zero when $2x = 0$, or $x - 2 = 0$, yielding $x = 0$ or $x = 2$. We indicate the number line in Figure 4.

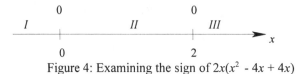

Figure 4: Examining the sign of $2x(x^2 - 4x + 4x)$

We choose values in each of the regions and determine the sign of the expression at these test values. They are indicated in Table 3.

Region	Test value	Sign of Product $2x(x-2)^2$
I	$x = -1$	$2(-1)(-3)^2 = -$
II	$x = 1$	$2(1)(-1)^2 = +$
III	$x = 3$	$2(3)(1)^2 = +$

Table 3: Sign of $2x(x-2)^2$

We now can complete the sign diagram as given in Figure 5.

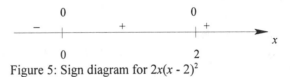

Figure 5: Sign diagram for $2x(x-2)^2$

The expression will be negative or zero when $x \leq 0$ or $x = 2$. We allow $x = 2$ because the inequality is $2x^3 - 8x^2 + 8x \leq 0$, allowing the expression to equal to zero. Using the interval notation we may write the solution as $(-\infty, 0)$ or $x = 2$. (The symbol $-\infty$ is read "minus infinity.")

(b) this reduces to solving the inequality $2x^3 - 8x^2 + 8x \geq 0$. Its solution is read immediately from Figure 4, namely the interval $x \geq 0$, or in interval notation, as $[0, \infty)$. (Note that when $x = 0$ or 2, the expression is zero, elsewhere in the interval it is positive.)

■■■

Example 2
Find those values of x so that $\dfrac{2x-5}{x+3} \leq 0$

Solution
The problem reduces to determining when the quotient on the left of the inequality is negative or zero. Since the rules of signs for quotients is the same as for products, (when the factors have the same sign the quotient will be positive, and when they are different it will be negative) we proceed in almost the identical way as above. The factors that determine the sign are the numerator and denominator. The numerator will be zero when $x = 5/2$ (verify!), and the denominator will be zero when $x = -3$. Be careful, x is not allowed to be -3, because then we would be dividing by zero. We indicate this on the number line by writing ND (not defined) above -3. See Figure 6.

Figure 6: Examining the sign of $\frac{2x-5}{x+3}$

We next choose and test points in each of the regions. They are indicated in Table 4.

Region	Test value	Sign of $\frac{2x-5}{x+3}$
I	$x = -4$	$-13/-1 = -/- = +$
II	$x = 0$	$-5/3 = -/+ = -$
III	$x = 4$	$3/7 = +/+ = +$

Table 4: Sign of $\frac{2x-5}{x+3}$

(Notice that all we really need is the sign not the actual value of the expression at the test values.) We now can complete the sign diagram as given in Figure 7.

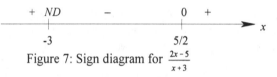

Figure 7: Sign diagram for $\frac{2x-5}{x+3}$

We may now read our solution directly from the sign diagram, namely $-3 < x \le 5/2$ or in interval notation, $(-3, \ 5/2]$. Note that $x \ne -3$, but $x = 5/2$ since the expression is not defined at -3, and is zero at 5/2.
■■■

Example 3
Solve the inequality $\dfrac{3x-8}{2x+5} > 1$.

Solution
We first rewrite the inequality so that one side is zero. We have $\dfrac{3x-8}{2x+5} - 1 > 0$. We combine into one fraction to obtain

$$\frac{x-13}{2x+5} > 0$$

The method of solution is now similar to the previous example. The numerator is zero when $x = 13$, and the denominator is zero (the fraction is not defined) at $x = -5/2$. We begin the sign analysis with Figure 8.

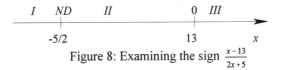

Figure 8: Examining the sign $\frac{x-13}{2x+5}$

To determine the sign in each region, we test points as indicated in Table 5.

Region	Test value	Sign of $\frac{x-13}{2x+5}$
I	$x = -3$	$-/- = +$
II	$x = 0$	$-/+ = -$
III	$x = 14$	$+/+ = +$

Table 5: Sign of $\frac{x-13}{2x+5}$

Thus, we have as our completed sign diagram,

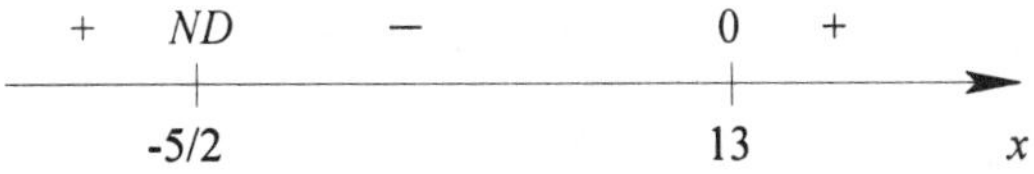

Figure 9: Sign Diagram for $\frac{x-13}{2x+5}$

This expression $\frac{x-13}{2x+5}$ is positive when $x < -5/2$ or $x > 13$ (which may also be written as $-\infty < x < -5/2$ or $13 < x < \infty$. Using interval notation, we may write the solution as $(-\infty, -5/2)$ or $(13, \infty)$.
∎∎∎

We remark that some text use the set theoretic symbol ∪ instead of writing the word "or." Thus, in place of $(-\infty, -5/2)$ or $(13, \infty)$ they would write $(-\infty, -5/2) \cup (13, \infty)$. Also, if an inequality has no solution we indicate this by writing ∅, which represents the empty set.

Our last example illustrates that as long as an expression is written in factored form, the procedure for finding the sign of an expression generalizes to any number of factors.

Example 4

Solve the inequality $\dfrac{4x^4(x^3 - 4x)}{(x^2 + 7x + 12)(x^2 - 9)} \geq 0$.

Solution
We must first factor the given expression.

$$\frac{4x^4(x^3-4x)}{(x^2+7x+12)(x^2-9)} = \frac{4x^4x(x^2-4)}{(x+3)(x+4)(x-3)(x+3)} = \frac{4x^5(x+2)(x-2)}{(x+4)(x+3)^2(x-3)}$$

Thus, we must solve the inequality

$$\frac{4x^5(x+2)(x-2)}{(x+4)(x+3)^2(x-3)} \geq 0$$

The numerator will be zero when $x = 0$, -2, or 2, and the denominator will be zero (the expression will not be defined) when $x = -4$, -3 or 3. We indicated these point on the number line in Figure 9.

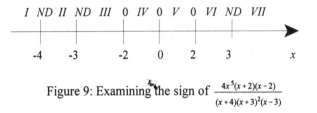

Figure 9: Examining the sign of $\frac{4x^5(x+2)(x-2)}{(x+4)(x+3)^2(x-3)}$

We need seven test points to determine the sign in each region. In determining the sign of the expression, rather than actually evaluating it for each test value, we will instead indicate the sign of each factor for each test value. All we need do is count the total number of negative signs, an odd number means the expression is negative, and even number means it is positive. Be careful when working with exponents. For example, $(-)^5$ counts as five negative signs, $(-)^2$ counts as two. Table 6 summarizes the sign of the expression $\frac{4x^5(x+2)(x-2)}{(x+4)(x+3)^2(x-3)}$ in each of the seven regions.

Region	Test value	Sign of $\frac{4x^5(x+2)(x-2)}{(x+4)(x+3)^2(x-3)}$
I	-5	$4(-)^5(-)(-)/(-)(-)^2(-) = -$
II	-3.5	$4(-)^5(-)(-)/(+)(-)^2(-) = +$
III	-2.5	$4(-)^5(-)(-)/(+)(+)^2(-) = +$
IV	-1	$4(-)^5(+)(-)/(+)(+)^2(-) = -$
V	1	$4(+)^5(+)(-)/(+)(+)^2(-) = +$
VI	2.5	$4(+)^5(+)(+)/(+)(+)^2(-) = -$
VII	4	$4(+)^5(+)(+)/(+)(+)^2(+) = +$

Table 6: Sign of $\frac{4x^5(x+2)(x-2)}{(x+4)(x+3)^2(x-3)}$

The information obtained from Table 5 is now placed into our sign diagram in Figure 10.

Figure 10: Sign of $\dfrac{4x^5(x+2)(x-2)}{(x+4)(x+3)^2(x-3)}$

We see that the expression will be zero or positive when $-4 < x < -3$ or $-3 < x \le -2$ or $0 \le x \le 2$ or $x > 3$. Using the interval notation, we can write it as $(-4, -3)$ or $(-3, -2]$ or $[0, 2]$ or $(3, \infty)$.

■■■

In each of the example we consider so far, the expression that needed to be analyzed were easily factorable. Even when this is not the case, we can sometimes perform the analysis needed to determine the sign of a given expression. The next example illustrates.

Example 5
Solve the inequality $x^2 + 2x - 2 < 0$.

Solution
The zeros of the quadratic $x^2 + 2x - 2$ are found by the quadratic formula to be $x_1 = -1 - \sqrt{3} \approx -2.732$, and $x_2 = -1 + \sqrt{3} \approx 0.732$. We now proceed in the usual way. We begin our sign diagram in Figure 11.

$$
\begin{array}{ccccc}
I & 0 & II & 0 & III \\
\end{array}
$$

$x_1 \approx -2.732 \qquad\qquad x_2 \approx 0.732$

Figure 11: Examining the sign of $x^2 + 2x - 2$

We next test a point in each region, as indicated in Table 7.

Region	Test value	Sign of $x^2 + 2x - 2$
I	$x = -3$	$9 - 6 - 2 = +$
II	$x = 0$	$-2 = -$
III	$x = 1$	$1 + 2 - 2 = +$

Table 7: Sign of $x^2 + 2x - 2$

138

We now complete the sign diagram as given in Figure 12.

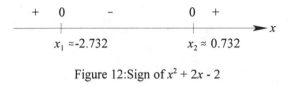

Figure 12:Sign of $x^2 + 2x - 2$

Therefore, we see that $x^2 + 2x - 2 < 0$ when $-1 - \sqrt{3} < x < -1 + \sqrt{3}$.

■■■

Suppose we have a quadratic whose zeros are complex numbers. That means there is no real number at which the quadratic can be zero, therefore the sign of the quadratic is always positive or always negative. We need only test any particular x-value to determine the sign of the quadratic. The next example illustrates.

Example 6
Solve the inequality $x^2 + 9 > 0$.

Solution
Since $x^2 + 9$ has no real zeros (verify!), then we need only determine its sign at any convenient x-value. We choose $x = 0$. We see that for this value of x the quadratic is positive, therefore it is everywhere positive, so we have as our solution the interval $-\infty < x < \infty$.

■■■

Exercise set 14
In each of the following exercises, solve the given inequality.

1. $2x - 3 > 0$

2. $2x - 3 \leq 0$

3. $(x - 3)(x + 4) \geq 0$

4. $(x - 3)(x + 4) < 0$

5. $(3x + 5)(4x - 7) \leq 0$

6. $(3x + 5)(4x - 7) > 0$

7. $(5x - 3)(x + 5) \geq 0$

8. $(5x - 3)(x + 5) < 0$

9. $(5 - 2x)(x + 4) > 0$

10. $(3 - 4x)(2 - 3x) \geq 0$

11. $x(2x - 5) \geq 0$

12. $x(2x - 5) < 0$

13. $(x + 2)^2(x - 3) \geq 0$

14. $(3x - 4)^4(2x - 3) \leq 0$

15. $x(x - 2)(x + 3) \leq 0$

16. $(x + 1)(x - 2)(x + 5) \geq 0$

17. $(2x + 3)(3x - 5)(x + 1) \leq 0$

18. $x^2 + 7x + 12 \geq 0$

19. $x^2 - 7x + 10 < 0$

20. $x^2 + x - 42 > 0$

21. $6x^2 - x - 12 \leq 0$

22. $6x^2 - x - 12 > 0$

23. $10x^2 + 7x - 12 \geq 0$

24. $10x^2 + 7x - 12 < 0$

25. $24x^2 + 10x - 25 \geq 0$

26. $24x^2 + 10x - 25 < 0$

27. $x^3 - 16x \geq 0$

28. $x^3 - 16x < 0$

29. $25x - x^3 \leq 0$

30. $25x - x^3 > 0$

31. $x^2 - 9 \leq 7$

32. $x^2 + 4 > 5$

33. $(2x + 3)(x - 3) \geq 11$

34. $(2x + 3)(x - 3) < 11$

35. $(3x - 2)(2x + 3) < 28$

36. $(3x - 2)(2x + 3) \geq 28$

37. $(4x - 5)(5x + 4) \geq -9$

38. $(4x - 5)(5x + 4) < -9$

39. $\dfrac{(x+2)(x-1)}{x-3} \leq 0$

40. $\dfrac{(x+2)(x-1)}{x-3} > 0$

41. $\dfrac{(2x-5)(3x-2)}{(x-4)^2} > 0$

42. $\dfrac{(2x-5)(3x-2)}{(x-4)^2} \leq 0$

43. $\dfrac{x(2x-3)^2}{(x-4)^3} \geq 0$

44. $\dfrac{x(2x-3)^2}{(x-4)^3} < 0$

45. $\dfrac{10}{x-3} < 2$

46. $\dfrac{10}{x-3} \geq 2$

47. $\dfrac{7x}{x+3} \geq 4$

48. $\dfrac{7x}{x+3} < 4$

49. $\dfrac{x}{x-1} \geq \dfrac{3}{2}$

50. $\dfrac{x}{x-1} < \dfrac{3}{2}$

51. $\dfrac{(x-1)(x+2)}{x-3} > 1$

52. $\dfrac{(x-1)(x+2)}{x-3} \leq 1$

53. $\dfrac{x^2-16}{x^2-25} \leq \dfrac{20}{11}$

54. $\dfrac{x^2-16}{x^2-25} > \dfrac{20}{11}$

55. $\dfrac{3x^2(x-2)(x+3)}{(3x+4)^4(2x-1)} \leq 0$

56. $\dfrac{3x^2(x-2)(x+3)}{(3x+4)^4(2x-1)} > 0$

57. $\dfrac{5x^2(4x-7)(2x+9)^3}{(x-2)^5(2x+1)^6} \leq 0$

58. $\dfrac{5x^2(4x-7)(2x+9)^3}{(x-2)^5(2x+1)^6} > 0$

59. $x^2 - 8 \leq 0$

60. $x^2 - 8 > 0$

61. $x^2 - 2x - 1 < 0$

62. $x^2 - 2x - 1 \geq 0$

63. $x^2 - 4x - 41 > 0$

64. $x^2 - 4x - 41 \leq 0$

65. $x^2 - 2x + 5 < 0$

66. $x^2 - 2x + 5 \geq 0$

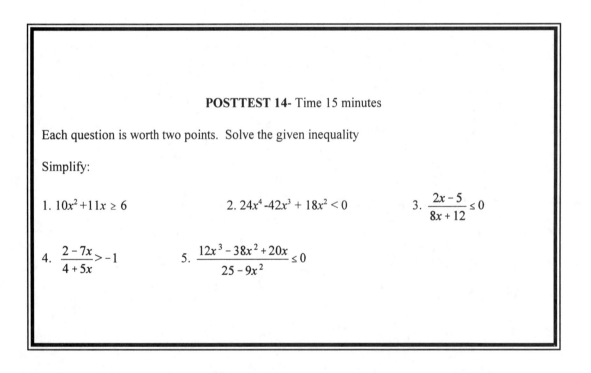

POSTTEST 14- Time 15 minutes

Each question is worth two points. Solve the given inequality

Simplify:

1. $10x^2 + 11x \geq 6$

2. $24x^4 - 42x^3 + 18x^2 < 0$

3. $\dfrac{2x - 5}{8x + 12} \leq 0$

4. $\dfrac{2 - 7x}{4 + 5x} > -1$

5. $\dfrac{12x^3 - 38x^2 + 20x}{25 - 9x^2} \leq 0$

15. The Line

» **Two Dimensional Coordinate System**
» **Horizontal and Vertical Lines**
» **The Slope Intercept Form**
» **Graphing**
» **The Point Slope Equation**
» **The Slope Formula**
» **An Economic Application**
» **The General Linear Equation**

PRETEST 15- Time 15 minutes

Each question is worth one point.

1. Determine the equation of the line whose slope is 2 and whose y-intercept is (0,5).

For questions 2- 5, use the line $5x - 2y = 9$.

2. Determine the slope of the line.

3. Determine its y-intercept.

4. Determine its x- intercept.

5. Find the equation of the line with slope 3 passing through the point (2, -5).

6. Find the equation of the line passing through (5, 2) and (5, -7).

7. Find the equation of the line passing through (5, 2) and (-3, 2).

8. Find the equation of the line passing through (4, 2) and (6, 7).

9. Find the of the equation line parallel to the line $3x - 2y = 7$ and passing through the point (6, 5).

10. Find the of the equation line perpendicular to the line $3x - 2y = 7$ and passing through the point (6, 5).

Coordinate geometry is one of the most useful tools in gaining a visual understanding of functions. With coordinate geometry, algebraic formulas may be translated into graphs. In many cases, having the graph is the end of the problem. As you know from everyday experience, a picture may be far more informative than a collection of data. In other cases, the picture may reveal the solution to a problem that might

142

otherwise appear to be too difficult to attack.

Although none of this should be new to you, let us review the ideas and thereby fix what will become our standard notation and terminology. We remark that our approach is probably not the one you first learned, but as a second time through the material, it will most quickly obtain the results needed to understand linear functions. We begin with a two-dimensional universe consisting of all ordered pairs of real numbers, usually denoted by $\boldsymbol{R^2}$. Examples of points in this universe are (1,4), (0.5 , 6), (0, $\sqrt{2}$), and (-π, 3.1). Because the two real numbers are ordered, (2,7) and (7,2) are different points. When coordinate axes are introduced in a plane, every ordered pair is associated with a point in the plane; and, conversely, every point in the plane has attached to it a unique ordered pair of coordinates. Let us briefly explain how this is done. First, as in Figure 1, a pair of number lines are drawn at right angles to one another, intersecting at the point zero on each line. The horizontal line is called the x-axis and the vertical line is called the y-axis. Construct a vertical line through any point in the plane. At the point where this line crosses the x-axis is a number called the x-coordinate (or abscissa) of the point. Now, construct a horizontal line through the point. At the point where this line crosses the y-axis is a number called the y-coordinate (or ordinate) of the point. If the point is called P, and its coordinates are (x, y), then we may refer to it as $P(x, y)$. In Figure 1, the points $P(2, 9)$, $Q(-4, -5)$, and $R(-5, 4)$, $S(4,-3)$ are shown. Notice that every point on the x-axis has ordinate 0 and every point on the y-axis has abscissa 0.

Two Dimensional Coordinate System

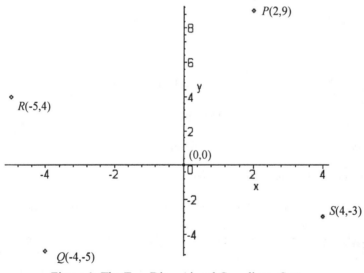

Figure 1: The Two Dimensional Coordinate System

The point where the axes intersect, called the *origin*, has coordinates (0,0). If we have a relation between two quantities x and y, then we may plot all the points whose coordinates satisfy the rule. The resulting picture is the graph of the relationship. Of course, there may be an infinite number of points, so that actually plotting them all is impossible. On the other hand, in most cases of practical interest, the graph assumes an easily observable pattern which we can visualize without seeing every point. It is also possible to go the other way. That is, we could have a verbal description of the geometric shape that we want as a graph and then try to find the algebraic relationship between the coordinates, that would produce it. For example, the vertical line crossing the x-axis at 3 would consist of all points with x-coordinate, 3. Thus, it would be described by the rule $x = 3$ and $y =$ anything. Since the last restriction is no restriction at all, we shall simply refer to this graph as "the line $x = 3$." In general, any vertical line will be represented by the equation $x = a$, where a is some constant. If a is positive, then the line lies to the right of the y-axis.

Horizontal and Vertical Lines

If a is negative, the line lies to the left of the y-axis. Of course, $x = 0$ is the y-axis. See Figure 2. We note that the graphs in Figure 2 cannot be graphs of functions of x, since there are many y values corresponding to the value $x = 3$. In a similar manner, the equation $y = b$ has as its graph a horizontal straight line which crosses the y-axis at $(0, b)$. If b is positive the line lies above the x-axis; if b is negative the line lies below the x-axis; and, naturally, $y = 0$ is the x-axis. See Figure 3 where we graph the lines $y = 4$ and $y = -3$. The function defined by $y = b$ is the simplest one there is and it is called the *constant function*.

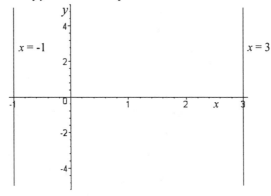

Figure 2: The lines $x = -1$ and $x = 3$

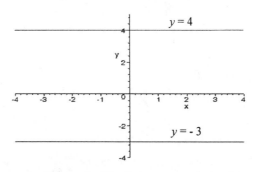

Figure 3 : The lines $y = 4$ and $y = -3$

Let us now see what kind of function gives rise to any other straight line graph. To begin, we consider a line passing through the origin and making an angle θ with the positive x-axis (see Figure 4). Let $P(x, y)$ be any point on the line, other than the origin. For any choice of (x, y), the triangles formed by the given line, the vertical line through P and the x-axis are similar. Therefore, from the figure, it is clear that the ratio y/x is the same for every point P. That is, y/x is a constant. We will denote the constant by m, and call this the *slope* of the line. That is, the coordinates of every point on this line except the origin must satisfy the condition that $y/x = m$, which is a constant. Multiplying by x, we obtain $y = mx$ and in this form we can also allow $(x, y) = (0, 0)$, so that $y = mx$

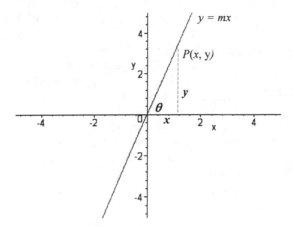

Figure 4: The line $y = mx$

is the equation of a line of slope m passing through the origin. Indeed, every non-vertical line passing through the origin is the graph of $y = mx$ for some value of m. Notice that for $m = 0$, this equation reduces to just $y = 0$, which we saw above is an equation for the x-axis.

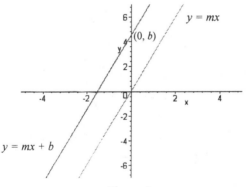

Figure 5

The Slope Intercept Form

Now consider the line parallel to the line $y = mx$ that cuts the y-axis at $(0,b)$. If b is positive, then this line lies above the first (as drawn in Figure 5) and for each x, the y-coordinate of every point on the new line is just b more than the value $y = mx$ on the first line. Therefore, we conclude that every point on this line satisfies the equation $y = mx + b$. Of course, if b is negative, the line $y = mx + b$ will lie below the original line, but in all other respects the analysis will be the same. *Note that parallel lines have the same slope.* Technically, the y-intercept is the point where the line crosses the y-axis and it has coordinates $(0, b)$. In practice, the number b will be referred to as the y-intercept. Thus, we conclude that the equation of any non-vertical straight line may be written in the *slope intercept form* which follows.

The Slope Intercept Form of a Line

An equation for any non-vertical line is

$$y = mx + b$$

where m and b are constants; m is called the slope of the line, and b is called the y-intercept.

Drawing the graph of a straight line is particularly simple, since we know that two points determine a line. Thus, we need locate only two points on the line and lay a straight edge across them. Let us look at some examples.

Graphing

Example 1.
Plot the graphs of   (a) $y = 2x - 1$ (b) $y = -\frac{1}{2}x + 4$ (c) $y = -3$
(d) $x = 3$

Solution.
In (a) we have a line of slope 2 and y-intercept -1. That is, the line crosses the y-axis at $(0, -1)$. In order to draw its graph we need only one additional point. So, we substitute any convenient value for x into the equation, say, $x = 1$. When $x = 1$, $y = 2(1) - 1 = 1$. That is, the line passes through $(0, -1)$ and $(1, 1)$. We plot these two points and draw the line as shown in Figure 6(a).

In (b), the slope is $-\frac{1}{2}$ and the y-intercept is 4. This line passes through $(0, 4)$. Picking $x = 2$ for convenience, we get a second point $y = -\frac{1}{2}(2) + 4 = 3$. Thus, for a second point we have $(2, 3)$. The graph is shown in Figure 6(b).

We see no x term in equation (c). However, we recognize this as being of the form $y = $ constant, which is a horizontal line consisting of all points with y-coordinate -3. Its graph is Figure 6(c). We note that this line could be thought of as $y = 0x + (-3)$, that is, a line of zero slope, parallel to the x-axis, with y-intercept -3.

Finally, in (d), there is no y term and we recognize this line as a special case that cannot be put into slope intercept form. Its graph is the vertical line shown in Figure 6(d).

■ ■ ■

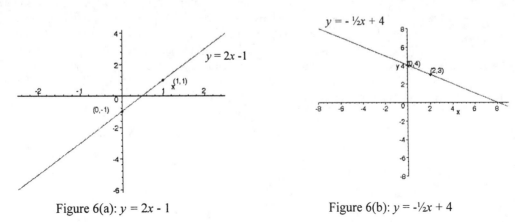

Figure 6(a): $y = 2x - 1$ Figure 6(b): $y = -\frac{1}{2}x + 4$

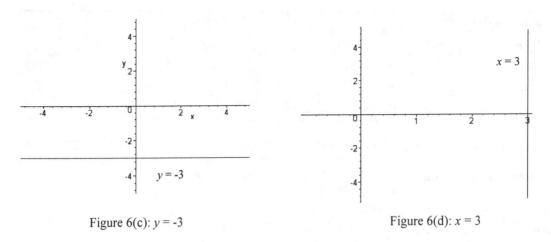

Figure 6(c): $y = -3$ Figure 6(d): $x = 3$

Example 2

Find an equation of a line having slope 3, and which passes through the point (-1, 4).

Solution.

Since this is not a vertical line, its equation can be written in slope intercept form

$$y = 3x + b,$$

where b is to be determined. Since the coordinates of the point (-1, 4) must satisfy the equation, we have

$$4 = 3(-1) + b$$

$$4 = -3 + b$$

Thus, $b = 7$, and the equation of the line is

$$y = 3x + 7$$

■■■

The Point-Slope Equation

In the many applications, we shall frequently encounter examples in which one knows the slope of a line and one point on the line as in Example 2. Therefore, it will be convenient to have a simple formula into which such data can be substituted to find the equation of the line directly. Let us suppose that we know the slope, m, and one point (x_1, y_1) on a line. As in the previous example, we know that the equation must be $y = mx + b$, where b is to be determined. Substituting the known point (x_1, y_1), yields

$$y_1 = mx_1 + b$$

We solve for b, obtaining

$$y_1 - mx_1 = b$$

We substitute this for b into $y = mx + b$, to obtain

$$y = mx + (y_1 - mx_1) = mx + y_1 - mx_1$$

Now we subtract y_1 from both sides of the equation, giving

$$y - y_1 = mx - mx_1.$$

Finally, we factor the m from both terms on the right hand side of the equation, to get

$$y - y_1 = m(x - x_1).$$

This is the so-called *point-slope* equation of a straight line.

The Point Slope Equation of a Line
If you are given the slope of a line m, and know the coordinates of one point (x_1, y_1) on the line, then you may determine an equation for the line by simply substituting the information into the point-slope equation

$$y - y_1 = m(x - x_1)$$

Example 3
Using the point-slope equation, rework Example 2.

Solution.
In Example 2, we were asked to find an equation for a line of slope 3, that passes through (-1, 4). The given information exactly suits the point-slope formula that was just derived. Therefore, we substitute directly

$$m = 3, x_1 = -1, y_1 = 4$$

to obtain

$$y - 4 = 3(x - (-1))$$

$$y - 4 = 3(x + 1)$$

If we want to, we can multiply out on the right hand side to get,

$$y - 4 = 3x + 3,$$

and adding 4 to both sides, we obtain

$$y = 3x + 7$$

the same form as before.
■■■

Here is another frequently encountered type of problem.

148

Example 4

(a) Find an equation for the line that passes through (2, 9) and (-5, -6). (b) Identify its slope and y-intercept.

Solution.

(a) Although we know more than one point, we cannot use the point-slope formula yet, since we do not know the slope. However, since both points satisfy the same equation, $y = mx + b$, we substitute each set of coordinates to get two equations

$$9 = 2m + b$$

$$-6 = -5m + b$$

If we subtract the lower equation from the upper, we will eliminate b, thus,

$$9 - (-6) = 2m - (-5m)$$

$$15 = 7m$$

$$m = 15/7$$

Now we may use the point-slope equation using either of the given points for (x_1, y_1). Let us use the first point (no reason to burden ourselves with extra negative signs):

$$y - 9 = (15/7)(x - 2)$$

Multiply through by 7 to simplify, yielding

$$7y - 63 = 15(x - 2)$$

$$7y - 63 = 15x - 30$$

$$7y = 15x + 33$$

(b) Notice that the equation we just obtained looks nice but it is not in the form $y = mx + b$ (or, if you prefer, $y = f(x)$). This is not unusual. Since we are asked to identify the slope and y-intercept of this line we must divide the equation by 7 to get

$$y = \frac{15}{7}x + \frac{33}{7}$$

Thus, the slope is 15/7 and the y-intercept is 33/7.

■■■

The Slope Formula

It is not uncommon to encounter cases in which two points are known, and an equation of the line determined by them is needed. What would help is a simple way to find the slope; then you can use the point-slope formula as in the last example. Therefore, let us suppose that we know the coordinates of two points on a non-vertical line. Call them $P(x_1, y_1)$ and $Q(x_2, y_2)$. Proceeding as in the preceding example, we realize that both points satisfy the equation

$$y = mx + b$$

So

$$y_2 = mx_2 + b$$

and

$$y_1 = mx_1 + b$$

subtracting the lower equation from the upper, we have

$$y_2 - y_1 = mx_2 - mx_1 = m(x_2 - x_1).$$

Since the line is not vertical, no two points on it have the same x-coordinate. Therefore, $x_1 \neq x_2$ and $x_2 - x_1 \neq 0$. Thus, dividing by $(x_2 - x_1)$, we obtain the *slope formula*.

The Slope Formula
Let (x_1, y_1) and (x_2, y_2) be any two points on a line for which $x_1 \neq x_2$, then the slope of the line is given by

$$m = \frac{y_2 - y_1}{x_2 - x_1}$$

This result is known as the *slope formula*. Do we have to worry about the possibility that $x_2 - x_1 = 0$? Not really. If the difference is zero, this means that $x_2 = x_1$. If the two points are different points and $x_2 = x_1$ then our line passes through two points with the same x-coordinate. That is, the line is vertical. The slope of a vertical line is undefined in any event, so we can safely use the above formula, and if the denominator is zero, we will know that the desired line is vertical. Note that it is correct to write

$$m = \frac{y_1 - y_2}{x_1 - x_2}$$

but

$$m \neq \frac{y_2 - y_1}{x_1 - x_2} \quad \text{and} \quad m \neq \frac{y_1 - y_2}{x_2 - x_1}$$

(Why?)

Example 5
Using the slope formula, rework Example 4.

Solution.
We know the two points, $(x_1, y_1) = (2, 9)$ and $(x_2, y_2) = (-5, -6)$. Substituting into the slope formula,

$$m = \frac{-6 - 9}{-5 - 2} = \frac{-15}{-7} = \frac{15}{7}$$

of course, now that we have m, we may finish the problem exactly as in Example 4.

■■■

150

The slope formula also gives us a good indication of what the slope really represents. Call the two given points, $P(x_1, y_1)$ and $Q(x_2, y_2)$ We can always think of Q as lying to the right of P (see Figure 7). That is, always identify x_2 and x_1, so that $x_2 > x_1$; then $x_2 - x_1$ is always positive. Thus, in the slope formula, the denominator is always positive. If Q is higher than P, the numerator is also positive; hence, m is positive. So, if the line is rising as you go from left to right, the slope is positive; otherwise it is negative. A zero slope indicates that the numerator is zero, which means that $y_2 = y_1$ and the line is horizontal.

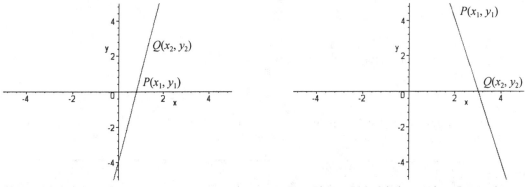

Figure 7(a) Q is higher than P, $m > 0$ Figure 7(a) Q is lower than P, $m < 0$

The numerator is the change in y as you go from the first point to the second, and the denominator is the change in x. Thus, m is sometimes referred to colloquially as the "rise" over the "run;" that is, the vertical change divided by the horizontal change. It now becomes clear that if the magnitude of the slope is large, then the change in y is large for a relatively small change in x. That is, steep lines have large slopes. Here, "large negative m" means $|m|$ is large. (Remember, $|m|$ means the *absolute value* or *magnitude* of m. So, for example, $|6| = 6$ and $|-6| = 6$.) Incidentally, the division between "large" and "small" in this context is 1. A line of slope 1 makes a 45° angle with the x-axis (-1 means the line makes an angle of 45° but measured from the "negative" portion of the x-axis.) Figure 8 shows several examples to give you an idea of how steep, lines of different slopes are..

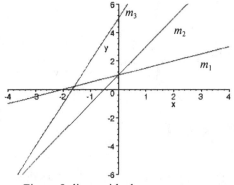

Figure 8: lines with slopes $m_1 < m_2 < m_3$

In summary, we have the following cases summarized in Table 1.

Slope	Property
$m < 0$	y decreases (gets smaller) as we move from left to right, i.e. the line slopes downhill.
$m > 0$	y increases (gets larger) as we move from left to right, i.e. the line slopes uphill.
$m = 0$	line is horizontal (parallel to x-axis).
undefined	line is vertical.

Table 1 - The relationship between a Line and its Slope

In short, we have seen equations of a straight line may take any of three forms:

1. $x = a$, a vertical straight line.,

2. $y = b$, a horizontal straight line.

3. $y = mx + b$, any other straight line.

The General Linear Equation

Of course, the second form of a line is only a special case of Form 3 with $m = 0$. All this can be summarized in the following:

Every equation of the form $Ax + By = C$, A, B, and C constants, A and B not both zero, is an equation of a straight line. Accordingly, every such equation is called a linear equation.

This may be verified as follows. Since A and B cannot both be zero, first consider the case where $B = 0$, and A is nonzero. The equation is now $Ax = C$, and we can divide by A yielding $x = C/A$. That is, $x =$ constant, which is a vertical line. On the other hand, if $B \neq 0$, then we can solve for B,

$$By = -Ax + C$$

Now divide by B,

$$y = -(A/B)x + (C/B)$$

which is of the form $y = mx + b$. Note that the slope of the line $Ax + By = C$ is $m = -A/B$.

∎

Example 6

Find the slope and y-intercept of the line whose equation is

$$3x + 4y = 8.$$

Also find the *x*-intercept (the point where the line crosses the *x*-axis) and plot the line.

Solution
The equation is $3x + 4y = 8$, which we solve for *y* in order to get it into the usual form:

$$4y = -3x + 8$$

Dividing by 4,

$$y = (-3/4)x + 2.$$

Now, by inspection, we see that the slope is -3/4 and the *y*-intercept is (0, 2). To find the *x*-intercept, we are really asking to find the value of *x* for which $y = 0$. In other words, we substitute $y = 0$ into the given equation and solve for *x*.

$$3x + 4(0) = 8$$

$$3x = 8$$

$$x = 8/3$$

The *x*-intercept is (8/3, 0). In general, unless a line passes through the origin, (or is horizontal or vertical) *the easiest way to draw its graph is to plot the intercepts*. Thus, in Figure 9 we show the intercepts and the line.

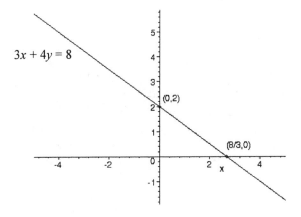

Figure 9: The Line $3x + 4y = 8$

■■■

Example 7
Determine an equation for the line parallel to the line $3x - 2y = 8$ and passing through the point (1,-2).

Solution
As above, we put the given line in the usual form by solving for *y*. We find that

$$y = 3/2x - 4.$$

Since the required line is parallel to the given line, its slope must also be 3/2. Thus, the required line has slope $m = 3/2$, passes through the point (1, -2), and by the point-slope formula its equation is

$$y - (-2) = \tfrac{3}{2}(x - 1)$$

Simplifying, we find that the slope-intercept equation of the required line is

$$y = 3/2x - 7/2.$$

We can also write an equation for the line in the form

$$3x - 2y = 7.$$

■■■

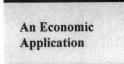

The slope is defined as the change in y-values divided by a corresponding change in x-values. Thus, slope is essentially the average rate of change of y with respect to x. It is precisely this interpretation of slope that is essential to our understanding of may applications, especially in Economics and Finance. The next example illustrates such an application.

Example 8

When a wholesaler sold 60 CD players at $60 per player, weekly sales averaged 150 players. For each $5 drop in the wholesale price the average number of players sold increased by 15. (a) Describe the relationship between the wholesale price and average weekly sales and (b) what is the average weekly sales if the wholesaler charges $42 per unit?

Solution

We let x represent the average weekly sales, and y the wholesale price.

(a) We first plot some points to see if we notice a pattern. When $y = 60$, $x = 150$, that is, (150, 60) is our first point. If we decrease the price, y by 5, then weekly sales, x increases by 15, so we have the point (165, 55), if we drop y again by 5, then x increases by another 15, and we have as our third point, (180, 50). These points are plotted in Figure 9.

Note that as we move to the right 15 units, we then fall 5 units to go from the point (150, 60) to the point (165, 55). Once again, as we move from the point (165, 55) and move right 15 units, we then fall another 5 units to the point (180, 50) and this trend continues. This ratio of the price to the weekly sales, or equivalently, the average rate of change of price with respect to weekly sales is the slope of the line connecting the points. Thus, we see a *linear* relationship between the variables y and x, the price and sales. The rate of change of price with respect to sales is 5/15 =1/3. Since sales are increasing as the price falls, we have a negative rate of change, that is, the slope is -1/3. We can now write the equation that represents this relationship. We have $m = -1/3$ and choosing the point (150, 60) we have

$$y - 60 = -1/3(x - 150)$$

or simplifying, we have

$$3y + x = 330$$

Since it makes sense to think of the price as driving the demand for the CDs, we solve for x as a function

154

of y. This gives

$$x = -3y + 330$$

(b) When the wholesale price $y = 42$, we have $x = -3(42) + 330 = 204$. Thus, at a price of $2 per units, the average sales are 204.

■■■

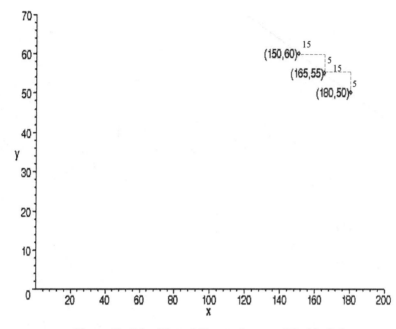

Figure 10: Price Plotted Versus Average Weekly Sales

We have already seen that parallel lines have the same slope, and conversely that two lines with the same slope are parallel. We now investigate the relationship between the slope of perpendicular lines, that is, lines that intersect at an angle of 90 degrees — right angles. Let us first dispense with the case where one line is vertical, then any line perpendicular to it must be horizontal. Thus, we assume in what follows that neither line is vertical. Without loss of generality, assume the two lines intersect at the origin. (Or equivalently, think of the origin as being at the intersection of the two lines.) Let m_1 and m_2 be the slope of these two lines, then their equations are $y = m_1 x$ and $y = m_2 x$ (why?). Choose points A and B on each of these lines with x-coordinate 1, then the corresponding y-coordinates are m_1 and m_2. Consider Figure 11.

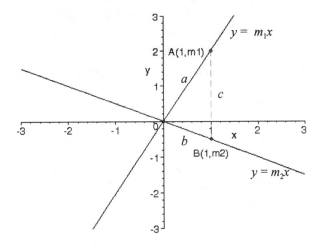

Figure 11: Slopes and Perpendicular Lines

Let us first assume the lines intersect at right angles at the origin. The triangle shown with sides a, b, and c is therefore a right triangle, and Pythagoras' theorem is applicable. By the distance formula, we have

$$a = \sqrt{(1-0)^2 + (m_1 - 0)^2} = \sqrt{1 + m_1^2}$$

$$b = \sqrt{(1-0)^2 + (m_2 - 0)^2} = \sqrt{1 + m_2^2}$$

and the vertical line connecting A to B has distance

$$c = m_1 - m_2$$

by Pythagoras' theorem

$$a^2 + b^2 = c^2$$

we have

$$1 + m_1^2 + 1 + m_2^2 = (m_1 - m_2)^2$$

or

$$2 + m_1^2 + m_2^2 = m_1^2 - 2m_1 m_2 + m_2^2$$

or

$$2 = -2m_1 m_2$$

or

$$m_1 m_2 = -1$$

Thus, we have shown that if two lines are perpendicular, the product of their slopes -1, or equivalently one is the negative reciprocal of the other, that is,

$$m_2 = -1/m_1$$

We also observe that each step in the above argument is reversible, that is, beginning with two lines whose slope product $m_1m_2 = -1$, by working backwards, we obtain the first statement which is Pythagoras' theorem, implying the triangle is a right triangle, therefore the lines intersect at right angles.

∎

Example 9
Determine the slope of the line perpendicular to the line (a) $y = 3x$ (b) $y = -2/5x + 7$ (c) $5x + 2y = 11$ (d) $x = 2$ (e) $y = 1$.

Solution
(a) $m_1 = 3$ therefore the slope of the perpendicular line is its negative reciprocal, $m_2 = -1/3$. (b)$m_1 = -2/5$ therefore its negative reciprocal is $m_2 = 5/2$. (c) rewriting the line in the form $y = -5/2x + 11/2$, we see its slope $m_1 = -5/2$ therefore the slope of the line perpendicular to it is $2/5$. (d) $x = 2$ is a vertical line, therefore any line perpendicular to it is horizontal. Horizontal lines have 0 slope. (e) The line $y = 1$ is a horizontal line, any line perpendicular to it is vertical and vertical lines have no slope.

∎∎∎

Example 10
Determine the equation of the line passing through the point (2, -3) and perpendicular to the line $3x - 2y = 12$.

Solution
We first find the slope of the line $3x - 2y = 12$, by rewriting it as $y = 3/2x - 6$, we see immediately that its slope $m_1 = 3/2$. The slope of the line perpendicular to it is the negative reciprocal so we have $m_2 = -2/3$. By the point slope formula, the equation of the required line is
$$y - (-3) = -2/3(x - 2)$$
or
$$y = -2/3x - 5/3$$
or
$$2x + 3y = -5$$

∎∎∎

Exercise set 15
In exercises 1 - 5, determine the equation of the line whose slope and y-intercept are given.

1.	2, (0, 4)	6.	$3x - 4y = 12$
2.	-3, (0, 0)	7.	$2x + 4y - 9 = 0$
3.	0, (0, -2)	8.	$4y = 5x + 12$
4.	½, (0, 1/4);	9.	$1.3x + 4.7y + 11.2 = 0$
5.	1/4, (0, 3)	10.	$16 - 4y = 35$

In exercises 6 - 10, find the slope, x-intercept and y-intercept of the given line.

In exercises 11 - 15 give an equation of the line with the given slope and passing through the given point.

11.   -3, (2, 5)

12.   ½, (-2, 0)

13.   1/6, (0, 0)

14.   14, (-1, 0.3)

15.   4.1, (3, 0)

In exercises 16 - 20 find equation for the given line.

16.   A vertical line passing through (-1,8).

17.   A horizontal line passing through (2,-6).

18.   A line with intercepts (0,4) and (-2,0).

19.   A vertical line passing through (12,2).

20.   A horizontal line passing through (-3,8).

In exercises 21 - 40 plot the lines found in

21.   Exercise 1

22.   Exercise 2

23.   Exercise 3

24.   Exercise 4

25.   Exercise 5

26.   Exercise 6

27.   Exercise 7

28.   Exercise 8

29.   Exercise 9

30.   Exercise 10

31.   Exercise 11

32.   Exercise 12

33.   Exercise 13

34.   Exercise 14

35.   Exercise 15

36.   Exercise 16

37.   Exercise 17

38.   Exercise 18

39.   Exercise 19

40.   Exercise 20

41.   (a) Find the slope of the line whose equation is $2x - 5y = 6$. Find the x and y-intercepts of the line. Plot the line. (b) Find the equation of the line parallel to the line given in (a) and passing through (-1,7). Plot the line on the same set of axes.

42.   (a) Find the slope of the line whose equation is $3x + 7y + 42 = 0$. Find the x and y-intercepts of the line. Plot the line. (b) Find the equation of the line parallel to the line given in (a) and four units above it. Plot the line on the same set of axes.

43.   (a) Find an equation for the horizontal line passing through the y-intercept of the line in 41(a). (b) Find an equation for the vertical line passing through the x-intercept of the line in 41(b).

44.   (a) Find an equation for the horizontal line passing through the y-intercept of the line in 42(a). (b) Find an equation for the vertical line passing through the x-intercept of the line in 42(b).

45. Find the equations of two lines parallel to the line $y = -3$, and 4 units from it.

46. Find the equations of two lines parallel to $x = 2$, and 6 units from it.

In Exercises 47 - 58 find the slope of the line passing through each pair of points

47. $(1, -2)$ and $(1, -1.4)$.

48. $(2, -9)$ and $(12, 5)$.

49. $(-1/4, 2/5)$ and $(0, 0)$.

50. $(1, 4)$ and $(2, 4)$.

51. $(12, 16)$ and $(12, -73)$.

52. $(0, 3)$ and $(-6, 0)$.

53. $(½, -2)$ and $(1/4, -1/4)$.

54. $(0, -9)$ and $(1/2, 3)$.

55. $(-1/3, 2/3)$ and $(0, 0)$.

56. $(1, -5)$ and $(2, -5)$.

57. $(½, 16)$ and $(½, -73)$.

58. $(0, 4)$ and $(-7, 0)$.

In Exercise 59-70 find the equations of the lines and plot the lines from

59. Exercise 47.

60. Exercise 48.

61. Exercise 49.

62. Exercise 50.

63. Exercise 51.

64. Exercise 52

65. Exercise 53

66. Exercise 54.

67. Exercise 55.

68. Exercise 56.

69. Exercise 57.

70. Exercise 58.

71. Determine an equation for the line parallel to $y = 3x - 7$ and passing through the point $(1, -5)$.

72. Determine an equation for the line (a) parallel (b) perpendicular to $2x - 5y = 9$ and passing through the point $(-2, -4)$.

73. Determine an equation for the line (a) parallel (b) perpendicular to $3x + 7y = 11$ and passing through the point $(1, -3)$.

74. (a) Plot the line $4x + 6y + 12 = 0$. Find the area of the triangle formed by the line, the x-axis and the y-axis. (b) Repeat for $Ax + By + C = 0$ for A, B, C positive.

75. Find an equation of a line whose y-intercept is 4 and such that the area of the triangle formed by the line and the two axes is 20 square units.(Two possible answers.)

76. The area of a triangle formed by a line and the two axes is 40 and the slope of the line is -5. Find an equation for the line. (Two possible answers.)

77. (a) Find the length of the portion of the line $7x + 12y = 84$ that is cut off by the two axes. (b) Repeat for $Ax + By + C = 0$ for A, B, C positive.

78.   (a) Plot the points   (-1, -7), (4, 2) and (8,4). (b) Do they lie on the same line? (c) How can you tell *without* plotting?

79.   In 1990 the Massachusetts Non-Resident State Income Tax calls for a tax of 5% on earned income and 10% on unearned income. Suppose a person has total income of $40,000 of which amount x is earned. Find her tax, t, as a function of x.

80.   When the price for a color television is $240, the average monthly sales for this item at a department store is 450. For each $10 increase in price, the average monthly sales fall by 20 units. What is the average monthly sales if the price is increased to $400 per color television?

81.   When the price is $50 per radio, a producer will supply 100 radios each month for sale. For each $2 increase in price the producer will supply an additional 6 radios. How many radios are supplied if their per unit price is $72?

82.   Plot each of the following lines on the same set of axes. (a) $y = 2x$ (b) $y = 2(x - 3)$ (c) $y = 2(x + 3)$ (d) How are these lines related?

83.   Plot each of the following lines on the same set of axes. (a) $y = 2x$ (b) $y - 4 = 2x$ (c) $y + 4 = 2x$ (d) How are these lines related?

84.   Plot each of the following lines on the same set of axes. (a) $y = 2x$ (b) $y - 4 = 2(x - 3)$ (c) $y + 4 = 2(x - 3)$ (d) $y - 4 = 2(x + 3)$ (e) $y + 4 = 2(x + 3)$ (f) How are these lines related?

85.   In general, how are the lines $y = mx + b$ and $y - k = m(x - h) + b$ related? (*m, b, h,* and *k* are constants.)

POSTTEST 15- Time 15 minutes

Each question is worth one point.

1. Determine the equation of the line whose slope is 3 and whose y-intercept is (0,-3).

For questions 2- 5, use the line $2x + 7y = 12$.

2. Determine the slope of the line.

3. Determine its y-intercept.

4. Determine its x- intercept.

5. Find the equation of the line with slope 3/2 passing through the point (-8, 7).

6. Find the equation of the line passing through (-2, 6) and (3, 6).

7. Find the equation of the line passing through (5, 3) and (5, -8).

8. Find the equation of the line passing through (3, 4) and (-8, 5).

9. Find the equation of the line parallel to $2x + 5y = 9$ and passing through the point (10, -3).

10. Find the equation of the line perpendicular to $2x + 5y = 9$ and passing through the point (10, -3).

16. The Circle

» **Definition of a Circle**
» **Equation of a Circle**
» **Graphing a Circle**
» **The Ellipse**

PRETEST 16 - Time 10 minutes

Each question is worth two points.

1. Determine the equation of the circle centered at (2, -3) with radius 4.

2. Determine the radius and center of the circle $(x + 2)^2 + (y - 3)^2 = 9$.

3. Determine the radius and center of the circle $x^2 + y^2 - 10x + 6y + 9 = 0$.

4. Sketch the graph of the circle $(x + 2)^2 + (y - 3)^2 = 9$.

5. Identify and sketch the graph whose equation is $16x^2 + 9y^2 = 144$.

Suppose someone asks you to draw, as accurately as you can, a circle on a piece of cardboard. How would you go about doing it? One approach is to attach a pencil to one end of a taut string and hold the other end fixed. Keeping the string taut move the pencil. As you are doing so, you are drawing an arc of a circle. Continuing once around, the circle is drawn.

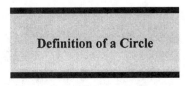
Definition of a Circle

The above method for sketching a circle may be used to define it. The circle is defined to be the set of all point equidistant from a given point. The given point is called the *center* of the circle and the distance each point is from the center is called the *radius*. This definition may be used to determine the equation of circle if we are given the coordinates of the center and the length of the radius. We illustrate in the following example.

Example 1
Find the equation of the circle center at the point C (1, 2) with radius 3.

Solution
Let P (x, y) be any point on the circle, our objective is to determine an equation relating x to y. We know that every point on the circle is 3 units from the center, therefore, we have the distance from the center

162

C (1, 2) to the point P (x, y) using the distance formula is

$$\sqrt{(x-1)^2+(y-2)^2} = 3$$

If we square both sides of the equation we obtain

$$(x-1)^2 + (y-2)^2 = 9$$

This is one way we can leave the equation, or if we like, we could multiply out and combine like terms and write the equation in the form

$$x^2 + y^2 - 2x + 4y - 4 = 0$$

■■■

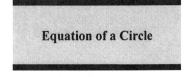
Equation of a Circle

What we did in Example 1, we may do in general to obtain the equation of the circle centered at the point C (h, k) with radius r. If the point $P(x, y)$ is any point on the circle, then it is a distance r from the center. Therefore, by the distance formula, we have

$$\sqrt{(x-h)^2+(y-k)^2} = r$$

or squaring both sides we have

$$(x - h)^2 + (y - k)^2 = r^2 \qquad\qquad (1)$$

Equation (1) is called the *standard* form and the most convenient form of the equation of a circle; inspection immediately yields the center and radius. However, if equation (1) is expanded, we obtain

$$x^2 + y^2 - 2hx - 2ky + h^2 + k^2 - r^2 = 0$$

If we let $a = -2h$, $b = -2k$ and $c = h^2 + k^2 - r^2$, then we may write the equation in the form

$$x^2 + y^2 + ax + by + c = 0 \qquad\qquad (2)$$

Equation (2) is called the *general* form, and is not the most useful form for the equation of the circle, as in is not immediately obvious that it is indeed a circle (as we shall see), and if it is, it is not immediately apparent where its center is located nor what is its radius.

When the circle is centered at the origin, that is, $h = k = 0$, equation (1) reduces to

$$x^2 + y^2 = r^2 \qquad\qquad (3)$$

The only difference between the circle whose equation is given by (1) and the one given by (3) is their location. Each point in the circle given by equation (1) is translated horizontally by h units and vertically by *k* units. We illustrate this remark in Example 5 below.

Example 2
Given the equation of the circle $(x - 2)^2 + (y - 4)^2 = 16$, determine is center and its radius.

Solution
Comparing with (1) we have immediately $h = 2$, $k = 4$, and $r^2 = 16$. Thus, the center of the circle is $C(2, 4)$ and the radius is $r = \sqrt{16} = 4$.
■■■

Looking at (1), we see that *h* is the value of *x* that makes the first parenthesis zero and *k* is the value of *y* that makes the second parenthesis vanish, thus, to find the *x*-coordinate of the center we may set $x - h = 0$ and to find the *y*-coordinate we may set $y - k = 0$.

Example 3
Given the equation of the circle $(x - 3)^2 + (y + 5)^2 = 25$, determine is center and its radius.

Solution
Using the observation made above, set $x - 3 = 0$ and $y + 5 = 0$, yielding $x = 3$ and $y = -5$. Thus, the center of the circle is $C(3, -5)$. The radius is $r^2 = 25$ or $r = 5$.
■■■

Graphing a Circle

Given the equation of a circle we can sketch its graph fairly easily. Actually, three points uniquely determine a circle, but we suggest the following approach which uses four points in addition to the center.

Graphing a Circle
1. Plot the center

2. Move horizontally to the right from the center a distance equal to the radius and plot this point.

3. Move horizontally to the left from the center a distance equal to the radius and plot this point.

4. Move vertically above the center a distance equal to the radius and plot this point.

5. Move vertically below the center a distance equal to the radius and plot this point.

6. Connect these four points with a smooth curve.

We illustrate this approach with the following example.

Example 3
Sketch the circle whose equation is given in Example 2.

Solution
The circle is centered at $C(2, 4)$ and has radius 4. We choose the four points suggested above. Moving

164

to the right 4 units from (2, 4) yields (6, 4). Moving to the left 4 units yields (-2, 4). Moving up 4 units yields (2, 8) and moving below 4 units yields (2, 0). The points are plotted and the circle is drawn in Figure 1.

■■■

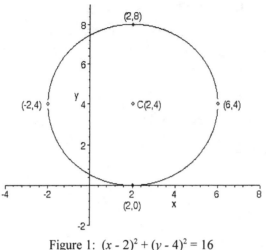

Figure 1: $(x - 2)^2 + (y - 4)^2 = 16$

When the general form is given for the equation of the circle we must first transform it into the standard form so we may easily locate its center and radius. In order to do this, we must complete the square, but in this case, twice; once for the x terms and once for the y terms. We must also remember that whatever we do to one side of an equation we must do to the other. We illustrate the procedure in the next example.

Example 4
Given the circle whose equation is $x^2 + y^2 - 6x + 10y + 9 = 0$. (a) Determine its center and radius. (b) Sketch its graph.

Solution
(a) We rewrite the equation as

$$x^2 - 6x + y^2 + 10y = -9$$

We first complete the square for the x terms: $\frac{1}{2}(- 6) = -3$, and $(-3)^2 = 9$, so we have

$$x^2 - 6x + \mathbf{9} + y^2 + 10y = -9 + \mathbf{9}$$

or

$$(x - 3)^2 + y^2 + 10y = 0$$

We next complete the square for the y terms: $\frac{1}{2}(10) = 5$ and $(5)^2 = 25$, so we have

$$(x - 3)^2 + y^2 + 10y + \mathbf{25} = 0 + \mathbf{25}$$

or

$$(x - 3)^2 + (y + 5)^2 = 25$$

This is the circle given in Example 3. We see that the circle is centered at (3, -5) and has radius 5.

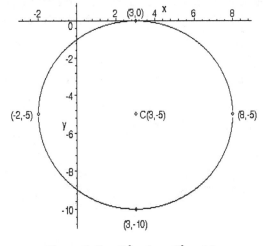

Figure 2: $(x - 3)^2 + (y + 5)^2 = 25$

(b) we plot its graph by using the four points described above. Moving to the right from the center 5 units we have the point (8, -5), moving to the left from the center we have (-2, -5). Moving up 5 units from the center we have (3, 0) and moving 5 units down from the center we have (3, -10). We plot these points and the graph in Figure 2.

■■■

Exercise 5
Sketch the circles $(x - 1)^2 + (y - 2)^2 = 9$ and $x^2 + y^2 = 9$ on the same coordinate system.

Solution
The sketch of the two circles is given in Figure 3, and the point used to plot the graphs are indicated.

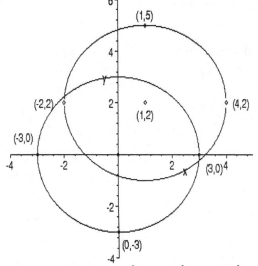

Figure 3: The graphs of $(x - 1)^2 + (y - 2)^2 = 9$ and $x^2 + y^2 = 9$

If we moved the circle centered at the origin so that its center would be at $(1, 2)$ then the two graphs would be coincident. That is, if any point on the circle centered at $(0,0)$ is moved one unit to the right and 2 units up then we get the corresponding point on the other circle. For example, take the point $(3, 0)$, $(3 + 1, 0 + 2)$ yields the corresponding point $(4, 2)$.

■■■

Not every equation of the form $x^2 + y^2 + ax + by + c = 0$ is a circle. Consider the next example.

Example 6
Classify each of the following: (a) $x^2 + y^2 + 4x - 6y + 13 = 0$, (b) $x^2 + y^2 + 4x - 6y + 15 = 0$

Solution
We complete the square to understand what is happening here.
(a) $x^2 + y^2 + 4x - 6y + 13 = 0$ is rewritten as

$$x^2 + 4x + y^2 - 6y = -13$$

$\frac{1}{2}(4) = 2$ and $2^2 = 4$, so we have
$$x^2 + 4x + \mathbf{4} + y^2 - 6y = -13 + 4$$

or

$$(x + 2)^2 + y^2 - 6y = -9$$

$\frac{1}{2}(-6) = -3$, and $(-3)^2 = 9$, so we have

$$(x + 2)^2 + y^2 - 6y + \mathbf{9} = -9 + \mathbf{9}$$
or
$$(x + 2)^2 + (y - 3)^2 = 0$$

The only way a sum of squares can be zero is if each square term is zero, therefore we have $x = -2$ and $y = 3$. That is, the equation reduces to a single point, namely the point $(-2, 3)$, (or the circle or radius zero centered at $(-2,3)$.)

(b) $x^2 + y^2 + 4x - 6y + 15 = 0$ is rewritten as

$$x^2 + 4x + y^2 - 6y = -15$$

$\frac{1}{2}(4) = 2$ and $2^2 = 4$, so we have
$$x^2 + 4x + \mathbf{4} + y^2 - 6y = -15 + 4$$

or

$$(x + 2)^2 + y^2 - 6y = -11$$

$\frac{1}{2}(-6) = -3$, and $(-3)^2 = 9$, so we have

$$(x + 2)^2 + y^2 - 6y + \mathbf{9} = -11 + \mathbf{9}$$
or
$$(x + 2)^2 + (y - 3)^2 = -2$$

It is impossible for the sum of squares to total a negative number. The sum must be either zero or a positive number. Therefore the equation given defines neither a curve nor a point, it results in a contradiction,

■■■

The last example indicates that the equation $x^2 + y^2 + ax + by + c = 0$ can either be a circle, a point or a contradiction.

Note that a circle does not describe a function. This may be seen at once from the vertical line test. Drawing a vertical line through the circle intersects the circle at two points with different y-values. Consider the circle centered at the origin with radius 3, $x^2 + y^2 = 9$ solving for y we have

$$y^2 = 9 - x^2$$

or

$$y = \pm\sqrt{9 - x^2}$$

(Note that to each x-value there corresponds two y-values.)

However, suppose we consider $y = \sqrt{9 - x^2}$, this is the upper half of the circle and to each x-value their corresponds one y-value. This half of the circle does indeed define a function. Its graph is given in Figure 4.

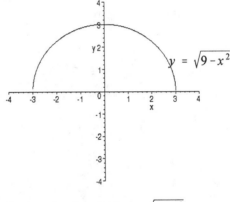

Figure 4: The Graph of $y = \sqrt{9 - x^2}$

The graph of $y = -\sqrt{9 - x^2}$ also defines a function. Its graph is given in Figure 5. Thus, a circle does not define a function, but its upper or lower halves taken separately do.

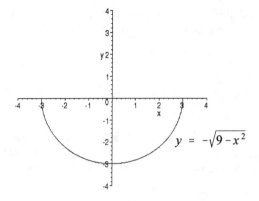

Figure 5: The Graph of $y = -\sqrt{9-x^2}$

The Ellipse

A circle centered at the origin has the equation $x^2 + y^2 = r^2$. What happens to the graph if we change the coefficients of the squared terms so that they are not the same? For example, let us consider the graph of the equation $4x^2 + 9y^2 = 36$. Such a graph is called an *ellipse*. It can be shown that multiplying the squared terms by constants scales the equation and the graph is no longer circular, but it is still closed. In fact, its graph is given in Figure 6.

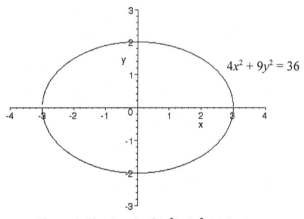

Figure 6: The Graph of $4x^2 + 9y^2 = 36$

The graph is obtained by finding the four intercepts. By setting $y = 0$, we obtain $4x^2 = 36$, and solving we have $x = -3$ or 3, thus the points (-3, 0) and (3, 0). Setting $x = 0$ we obtain $9y^2 = 36$ or $y = -2$ or 2, yielding the points (0, -2) and (0, 2).

More generally, any equation of the form $\dfrac{x^2}{a^2} + \dfrac{y^2}{b^2} = 1$ is the equation of an ellipse centered at the origin.

If the fractions are cleared, this equation has the form $Ax^2 + By^2 = C$. It is most easily drawn by finding the x and y intercepts and plotting them. The equation $4x^2 + 9y^2 = 36$ may be rewritten in the standard form by dividing by 36 to obtain $\dfrac{x^2}{9} + \dfrac{y^2}{4} = 1$.

Example 7

Sketch the graph of the ellipse $\dfrac{x^2}{25} + \dfrac{y^2}{9} = 1$.

Solution

Note that multiplying by the *LCD* which is 225, we could rewrite this equation as $9x^2 + 25y^2 = 225$. To find the x-intercepts, we set $y = 0$ and obtain $9x^2 = 225$, or $x^2 = 25$, solving, $x = -5$ or 5, thus the x-intercepts are $(-5, 0)$ and $(5, 0)$. To find the y-intercepts, we set $x = 0$ and obtain $25y^2 = 225$, or $y^2 = 9$, solving, $y = -3$ or 3, thus the y-intercepts are $(0, -3)$ and $(0, 3)$. The graph is plotted in Figure 7.

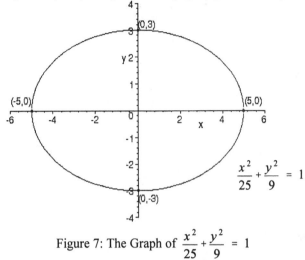

Figure 7: The Graph of $\dfrac{x^2}{25} + \dfrac{y^2}{9} = 1$

■■■

When we examined the graphs of the two circles $x^2 + y^2 = r^2$ and $(x - h)^2 + (y - k)^2 = r^2$, we saw that they were identical except the first one is centered at the origin and the second is centered at (h, k). Now consider the graphs of $\dfrac{x^2}{25} + \dfrac{y^2}{9} = 1$ and $\dfrac{(x-1)^2}{25} + \dfrac{(y-2)^2}{9} = 1$, they are also identical except that the second one has its center at the point $(1, 2)$, see Figure 9. Thus, in general $\dfrac{(x-h)^2}{a^2} + \dfrac{(y-k)^2}{b^2} = 1$ is identical to $\dfrac{x^2}{a^2} + \dfrac{y^2}{b^2} = 1$ except its center (and all its other points) are moved h units in the x-direction and k units is the y-direction. We shall leave the analysis of these translated ellipses to the exercises.

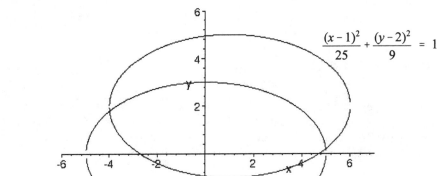

$$\frac{(x-1)^2}{25} + \frac{(y-2)^2}{9} = 1$$

$$\frac{x^2}{25} + \frac{y^2}{9} = 1$$

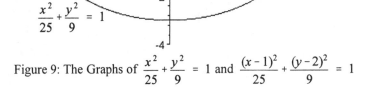

Figure 9: The Graphs of $\dfrac{x^2}{25} + \dfrac{y^2}{9} = 1$ and $\dfrac{(x-1)^2}{25} + \dfrac{(y-2)^2}{9} = 1$

Exercise set 16

In exercises 1 - 15 determine the center and radius of the given circle and sketch its graph.

1. $(x - 3)^2 + (y - 2)^2 = 4$

2. $(x - 4)^2 + (y - 2)^2 = 9$

3. $(x + 4)^2 + (y - 3)^2 = 16$

4. $(x - 3)^2 + (y + 2)^2 = 25$

5. $(x + 1)^2 + (y + 4)^2 = 4$

6. $(x + 3)^2 + (y + 2)^2 = 9$

7. $x^2 + y^2 = 4$

8. $x^2 + y^2 = 25$

9. $x^2 + y^2 + 6x - 10y + 9 = 0$

10. $x^2 + y^2 - 2x + 4y + 1 = 0$

11. $x^2 + y^2 + 8x - 6y + 9 = 0$

12. $2x^2 + 2y^2 - 6x - 10y + 9 = 0$

13. $9x^2 + 9y^2 - 12x + 24y - 101 = 0$

14. $16x^2 + 16y^2 - 48x + 8y - 27 = 0$

15. $36x^2 + 36y^2 - 48x + 60y - 283 = 0$

In exercises 16 - 25 determine whether the given equations is a circle, a point, or a contradiction (no real graph).

16. $x^2 + y^2 + 1 = 0$

17. $(x - 1)^2 + (y + 2)^2 + 4 = 0$

18. $(x + 4)^2 + (y + 3)^2 + 9 = 0$

19. $(x + 4)^2 + (y - 3)^2 - 9 = 0$

20. $(x + 3)^2 + (y - 2)^2 = 0$

21. $(x - 3)^2 + (y + 2)^2 = 0$

22. $x^2 + y^2 - 2x + 4y + 12 = 0$

23. $x^2 + y^2 + 4x - 6y + 13 = 0$

24. $x^2 + y^2 - 10x + 4y + 29 = 0$

25. $x^2 + y^2 + 8x + 6y + 16 = 0$

26. Under what conditions on a, b, c, and d is $ax^2 + ay^2 + bx + cy + d = 0$ (a) a circle, (b) a point, (c) a line. (d) a contradiction?

27. Let $y = f(x)$ describe the upper half of the circle $x^2 + y^2 = 16$. Determine (a) $f(-2)$, (b) $f(0)$, (c) $f(2)$.

28. Let $y = g(x)$ describe the lower half of the circle $x^2 + y^2 = 16$. Determine (a) $g(-2)$, (b) $g(0)$, (c) $g(2)$.

29. Let $y = f(x)$ describe the upper half of the circle $x^2 + y^2 + 8x - 6y + 9 = 0$. Determine (a) $f(-3)$, (b) $f(-6)$, (c) $f(-1)$.

30. Let $y = g(x)$ describe the upper half of the circle $x^2 + y^2 + 8x - 6y + 9 = 0$. Determine (a) $g(-3)$, (b) $g(-6)$, (c) $g(-1)$.

31. Let $y = f(x)$ describe the upper half of the circle $x^2 + y^2 + 6x - 10y + 9 = 0$, and $y = g(x)$ the lower half of the circle. Determine (a) $f(-4)$, (b) $g(-4)$, (c) $f(1)$, (d) $g(1)$.

32. On the same coordinate axes, sketch the graph of:
    (a) $x^2 + y^2 = 4$
    (b) $(x - 2)^2 + (y + 3)^2 = 4$,
    (c) $(x + 1)^2 + (y - 4)^2 = 4$

    How are the graphs related.

In exercises 33 - 39 sketch the graph of the given ellipse, labeling all intercepts.

33. $\dfrac{x^2}{9} + \dfrac{y^2}{25} = 1$

34. $\dfrac{x^2}{25} + \dfrac{y^2}{9} = 1$

35. $\dfrac{x^2}{4} + \dfrac{y^2}{16} = 1$

36. $\dfrac{x^2}{16} + \dfrac{y^2}{4} = 1$

37. $9x^2 + 4y^2 = 36$

38. $4x^2 + 9y^2 = 36$

39. Let $y = f(x)$ denote the upper half of the ellipse $\dfrac{x^2}{25} + \dfrac{y^2}{9} = 1$, and $y = g(x)$ its lower half. Determine (a) $f(2)$, (b) $g(2)$, (c) $f(-1)$, (d) $g(-1)$.

40.   Sketch the graph of $\frac{(x-1)^2}{4} + \frac{(y+2)^2}{9} = 1$.

41.   Sketch the graph of $\frac{(x+4)^2}{16} + \frac{(y-3)^2}{4} = 1$

POSTTEST 16 - Time 10 minutes

Each question is worth two points.

1. Determine the equation of the circle centered at (-2, 3) with radius 5.

2. Determine the radius and center of the circle $(x - 4)^2 + (y + 3)^2 = 16$.

3. Determine the radius and center of the circle $x^2 + y^2 + 8x - 10y + 16 = 0$.

4. Sketch the graph of the circle $(x - 4)^2 + (y + 3)^2 = 16$.

5. Identify and sketch the graph whose equation is $9x^2 + 4y^2 = 36$.

17. Solving Two Equations in Two Unknowns

> » **Method of Elimination**
> » **Method of Substitution**

PRETEST 17- Time 15 minutes

Each question is worth two points. Solve the given system of equations for x and y.

1. $2x - 5y = -2$
   $3x + 4y = 4$

2. $5x - 7y = 9$
   $11x - 4y = 15$

3. $\dfrac{3}{4}x - \dfrac{2}{3}y = 5$

   $\dfrac{5}{6}x + \dfrac{3}{4}y = 3$

4. $x = 3 - 5y$
   $2x - 15y = 31$

5. $\dfrac{2}{x} + \dfrac{3}{y} = 5$

   $\dfrac{5}{x} + \dfrac{2}{y} = -3$

Two non-parallel lines have a unique point of intersection. One way to locate this point is to plot each of these lines and then from the graph, read off the coordinates at their intersection. Unfortunately, if the point is not at integer coordinates, the graphical solution is usually at best an approximation. However, the intersection point is easily found exactly by simple algebraic methods. We examine two such methods in this section.

Instead of asking to find the point of intersection of the two lines (a graphical interpretation) we ask instead to solve the system of linear equations, or to find their simultaneous solution.

The first of the two methods is often called the *Method of Elimination*. Some texts call it the addition-subtraction method, a name which is illustrative of the method. The procedure is straightforward: given

Method of Elimination

two linear equations multiply each equation by a non-zero number so that one of the coefficients of one unknown in one equation is equal and opposite to the coefficient of that unknown in the other equation. This gives an equivalent system of equations having the same solution.

Adding the modified equations together results in a linear equation in one unknown, whose solution is found using the methods in Section 8. We illustrate the method through examples.

Example 1
Solve the following system of equations for both x and y:
$$3x + 4y = 18$$
$$5x - 3y = 1$$

Solution
We choose an unknown that we wish to eliminate, the choice arbitrary, we choose x. The *LCM* (Least Common Multiple) of the coefficients of the x-terms, 3 and 5 is 15. Therefore we multiply the first equation by 5 and the second equation by -3. We used negative 3 so the resulting coefficients would be equal and opposite. We have the following equivalent system,

$$15x + 20y = 90$$
$$-15x + 9y = -3$$

Adding these equations together, we have
$$29y = 87$$

Solving this equation, we have $y = 3$.

Now that we have determined y, we may either substitute this value for y in any of the above equations, or begin the process anew and eliminate y to find x. We choose to substitute for y in the first equation. This gives

$$3x + 4(3) = 18$$

or

$$3x = 6$$

or

$$x = 2.$$

Thus, our solution is $(2, 3)$, that is $x = 2$ and $y = 3$. Of course, this solution must check **both** of the original equations, we leave it as an exercise for you to check the solution.
■■■

We remark that after the elimination is performed and one unknown is found, we may substitute in any of the equations to find the other unknown, or repeat the process and eliminate the other unknown. When the numbers are "nice," substitution is convenient, when the numbers are not as "nice," elimination is the easiest approach.

Example 2
Solve the following system of equations for both x and y:
$$3x + 3y = 11$$
$$6x - 11y = 15$$

Solution
We shall eliminate x from the equations. The *LCM* of 3 and 6 is 6, therefore we multiply the first equation

by -2 (remember, we want the signs to be opposite) and leave the second equation alone. This gives us the following equivalent system:

$$-6x - 6y = -22$$
$$6x - 11y = 15$$

Adding the equations together yields

$$-17y = -7$$

or

$$y = 7/17.$$

We could substitute this result for y in any of the above equations and then solve for x, or if we choose to avoid fractions, we could solve for x by eliminating y from the original set of equations. The *LCM* of 3 and 11 is 33; we multiply the first equation by 11 and the second by 3 to obtain the following equivalent system.

$$33x + 33y = 121$$
$$18x - 33y = 45$$

Adding these equations together we have

$$51x = 166$$

or

$$x = 166/51$$

Thus, our solution is (166/51, 7/17) or $x = 166/51$ and $y = 7/17$.

■■■

The above two examples illustrate the basic notions of the elimination method. We chose to always have the terms equal and opposite so we always add the modified equations. If instead the terms were equal, then you could just as well subtract the two equations, that is why the method is also known as the addition-subtraction method.

The elimination methods works on any linear system of equations. However, sometimes, it is more convenient to first rewrite the equations. We suggest that when using the elimination method, each equation be put in the form

$$Ax + By = C$$

Where A, B and C are integers.

Example 3

Solve the following system of equations for both x and y:

$$\frac{2}{3}x + \frac{3}{4}y = -19$$

$$\frac{5}{6}x - \frac{4}{9}y = 26$$

176

Solution

We first clear fractions in each of the equations. The *LCD* of the first equation is 12 and the *LCD* of the second equation is 18. We multiply each term in the first equation by 12 and each term in the second equation by 18 to obtain the following equivalent system.

$$8x + 9y = -228$$
$$15x - 8y = 468$$

We now have an equivalent system of equations containing integers. The *LCM* for 9 and 8 is 72, so we multiply the first equation by 8 and the second equation by 9 to obtain the following equivalent system

$$64x + 72y = -1824$$
$$135x - 72y = 4212$$

Adding the equations together yields
$$199x = 2388$$
or
$$x = 2388/199 = 12$$

Substituting into $8x + 9y = -228$ gives $8(12) + 9y = -228$, or $9y = -324$, or $y = -324/9 = -36$. Thus, we have as our solution (12, -36) or $x = 12$ and $y = -36$.

■■■

We illustrate one more example where the elimination method may be used to solve a special non-linear system. Consider the following example.

Example 4
$$\frac{3}{x} + \frac{4}{y} = 18$$
$$\frac{5}{x} - \frac{3}{y} = 1$$

Solution

Let $X = 1/x$ and $Y = 1/y$ (it then follows that $x = 1/X$ and $y = 1/Y$) . The system then becomes

$$3X + 4Y = 18$$
$$5X - 3Y = 1$$

This is identical with the system in Example 1, with X and Y replacing x and y respectively. Thus, from Example 1, we have $X = 2$ and $Y = 3$. Therefore, since $x = 1/X$, and $y = 1/Y$, we have, $x = \frac{1}{2}$ and $y = 1/3$.

■■■

Any linear system of equation can be solved with the elimination method. However, there is another very powerful method, which, in certain types of examples, is easier to apply and most importantly, is effective in finding the solution to non-linear systems, as illustrated in Section 18. This is the method of substitution. Given a system of equations in two unknowns,

Method of Substitution

select one variable in either of the equations, solve for it in terms of the other variable, and then substitute this result for the selected variable in the other equation, thereby eliminating the selected variable in the second equation. We illustrate the method in the following examples.

Example 5
Solve the following system of equations for both x and y:
$$y = 3x - 4$$
$$3x + 2y = 46$$

Solution
We observe that y is given in terms of x in the first equation, therefore we take this expression and substitute it for y in the second equation. Thus, the second equation becomes

$$3x + 2(3x - 4) = 46$$

or

$$9x - 8 = 46$$

$$9x = 54$$

$$x = 6.$$

We now find y by substitution, $y = 3(6) - 4 = 14$. Thus our solution is $(6, 14)$ or $x = 6$ and $y = 14$.

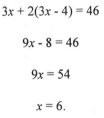

The previous example illustrates that once we have one unknown in terms of another, we can substitute to obtain an equation in one unknown, which is easily solved Thus, the substitution method is certainly convenient when we have, or it is easy to obtain, one of the unknown in terms of the other.

Example 6
Solve the following system of equations for both x and y:
$$7x - 2y = 11$$
$$x + 3y = -5$$

Solution
It is convenient to solve for x in the second equation, giving

$$x = -5 - 3y$$

We substitute this expression for x in the first equation to obtain

$$7(-5 - 3y) - 2y = 11$$

or

$$-23y - 35 = 11$$

or

$$-23y = 46$$

$$y = -2$$

Substituting for y, we have

$$x = -5 - 3(-2) = 1$$

Thus our solution is $(1, -2)$ or $x = 1$ and $y = -2$.

∎∎∎

The method of solution is most convenient when we have a simple expression for one of the unknowns in terms of the other. However, it can be used even when this is not the case. For example consider the system given in Example 1.

$$3x + 4y = 18$$
$$5x - 3y = 1$$

This system is most easily solved using the method of elimination, but suppose we choose to use the method of substitution. If we solve for y in the first equation (we chose this variable arbitrarily, any variable from either equation may be chosen), we have

$$y = \frac{18 - 3x}{4}$$

Substituting for y in the second equation gives

$$5x - 3\left(\frac{18 - 3x}{4}\right) = 1$$

We clear fractions by multiplying each term on both sides of the equation by 4 to obtain,

$$20x - 3(18 - 3x) = 4$$

or

$$29x - 54 = 4$$

$$29x = 58$$

$$x = 2$$

Substituting, we find $y = 3$, and our solution is $(2, 3)$. Clearly, in examples like this one, the method of elimination contains fewer steps.

We shall see, in Section 18, how the substitution method may be generalized to solve non-linear system of equations.

Exercise set 17

Using the method of elimination to solve the system of equation in exercises 1 - 12.

1. $2x + y = 7$
    $3x - 2y = 0$

2. $4x - 5y = -23$
    $2x + 3y = 5$

3. $5x - 3y = 35$
    $6x + 2y = 14$

4. $7x - 11y = 13$
$\quad 4x + 7y = 34$

5. $16x - 9y = 5$
$\quad 10x + 15y = 10$

6. $12x + 5y = 8$
$\quad 8x - 15y = 9$

7. $5x + 14y = 13$
$\quad 20x - 21y = -3$

8. $16x - 9y = 12$
$\quad -24x + 27y = -27$

9. $\dfrac{3}{4}x - \dfrac{2}{5}y = 10$
$\quad \dfrac{1}{2}x + \dfrac{3}{4}y = -\dfrac{7}{2}$

10. $\dfrac{2}{3}x + \dfrac{3}{4}y = 29$
$\quad \dfrac{3}{5}x - \dfrac{5}{6}y = 8$

11. $\dfrac{2}{x} + \dfrac{3}{y} = 12$
$\quad \dfrac{5}{x} - \dfrac{2}{y} = 11$

12. $\dfrac{4}{x} + \dfrac{5}{y} = -2$
$\quad \dfrac{3}{x} + \dfrac{7}{y} = 5$

Solve, using the method of substitution, the system equations in exercises 13- 20.

13. $y = 2x - 3$
$\quad -3x + 2y = 2$

14. $3x + y = -7$
$\quad 2x - 7y = 3$

15. $2x + 3y = 6$
$\quad x + 3y = 9$

16. $5x - 7y = 31$
$\quad x + 5y = -13$

17. $2x + 4y = 50$
$\quad 9x - 5y = 18$

18. $6x - 3y = 18$
$\quad 3x + 4y = 31$

19. $4x - 5y = -23$
$\quad 2x + 3y = 5$

20. $16x - 9y = 5$
$\quad 10x + 15y = 10$

POSTTTEST 17- Time 15 minutes

Each question is worth two points. Solve the given system of equations for x and y.

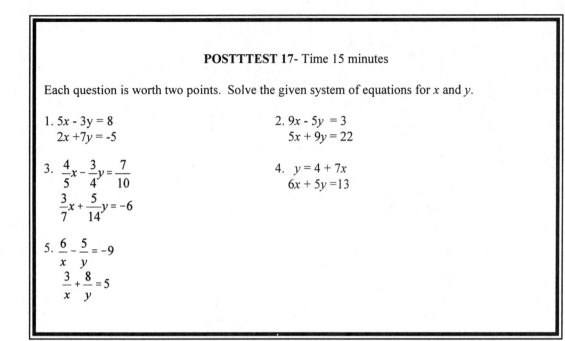

1. $5x - 3y = 8$
    $2x + 7y = -5$

2. $9x - 5y = 3$
    $5x + 9y = 22$

3. $\dfrac{4}{5}x - \dfrac{3}{4}y = \dfrac{7}{10}$

    $\dfrac{3}{7}x + \dfrac{5}{14}y = -6$

4. $y = 4 + 7x$
    $6x + 5y = 13$

5. $\dfrac{6}{x} - \dfrac{5}{y} = -9$

    $\dfrac{3}{x} + \dfrac{8}{y} = 5$

18. Non Linear Systems of Equations

» **Number of Solutions**
» **Method of Substitution**
» **Method of Elimination**
» **A Calculus Application**

PRETEST 18- Time 20 minutes

Each question is worth two points. Solve the given system of equations for the real values of x and y.

1. $2x + y = 7$
   $3x^2 - 3xy + y^2 = 61$

2. $2x^2 - 3y^2 = 5$
   $xy = 12$

3. $2x^2 - 5y^2 = 30$
   $3x^2 + 2y^2 = 83$

4. $y = 2x^2 + 5$
   $y^2 - 2x^2 = 47$

5. $x^2 + 2xy + 2y^2 = 10$
   $x^2 + 2y^2 = 6$

In the last section, when studying the solution of linear equations, you learned two methods, the method of elimination, and the method of substitution. In only very specialized problems does the elimination

---

Number of Solutions

method generalize, but the method of substitution proves to be especially valuable. We shall consider second order non-linear systems in two unknowns, that is, equations containing sums of terms involving x, y, x^2, y^2, and xy. For the most part, we shall consider systems that reduce to the solution of a quadratic equation. When solving such systems, we shall see that there may be none, one, two, three or four real solutions; this occurs because the graph of the second order non-linear equation is either a circle, ellipse, hyperbola or parabola (except in certain degenerate cases).

Consider the graphs in Figure 1. In Figure 1(a) the line is tangent to the circle at one point. In Figure 1(b) a line intersects a circle in two different points, the line is represented by a linear equation and the circle by a second degree equation. In Figure 1(c) we see the curve (a parabola) is tangent to a circle at its vertex and intersects the circle in two different points, each of the curves in Figure 1 (c) is represented by a second

182

degree equation. In Figure 1(d) we see the parabola intersects the circle in four different points, each of the curves in Figure(d) is represented by a second degree equation. It is also possible that the graphs do not intersect, meaning there is no real solution (only a complex solution). This last case is illustrated in Figure 1(e). In this section, we shall be mostly interested in real solutions, that is, the location of the points at which the graphs intersect.

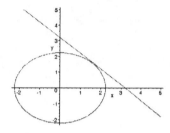 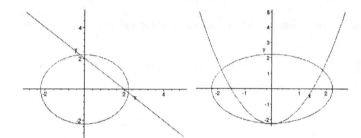

Figure 1(a)-One Intersection   Figure 1(b)-Two Intersections   Figure 1(c)-Three Intersections

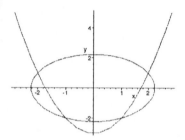 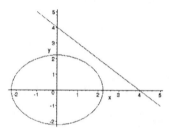

Figure 1(d)-Four Intersections   Figure 1(e)-No Intersections

Solving the system of equations corresponding to each of the above graphs yields the intersection points as their solution(s).

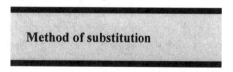

Method of substitution

The basic question is given a system of equations, how does one go about finding their (simultaneous) solution? If one of the equations is a linear equation, then generally, the most effective way to solve the system is to solve the linear equation for one of the variables in terms of the other, and then substitute for this variable in the remaining equation. We illustrate with examples. Note that we shall provide sketches of the graphs representing the equations, to illustrate the intersections, but we do not expect, at this time, for you to draw these sketches.

Example 1
Solve the following system for x and y.
$3x - y = 5$
$2x^2 - 2xy + y^2 = 5$

Solution

The first equation is linear, so we solve it for y and obtain

$$y = 3x - 5 \tag{1}$$

We next substitute for y in the second equation to obtain

$$2x^2 - 2x(3x - 5) + (3x - 5)^2 = 5$$

Multiplying out and combining terms, we obtain the quadratic equation

$$5x^2 - 20x + 20 = 0$$

This may be factored as

$$5(x - 2)^2 = 0$$

yielding

$$x = 2$$

Substituting for x in equation (1), we obtain $y = 3(2) - 5 = 1$. Thus, this non-linear system of equations has exactly one solution, namely $x = 2$ and $y = 1$, or $(2,1)$. The graph indicating these equations and their intersection is given in Figure 2.

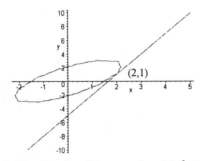

Figure 2: The Graphs of $3x - y = 5$ and $2x^2 - 2xy + y^2 = 5$

(Note that the oval shaped graph of the non-linear equation is an *ellipse*.)

■■■

Example 2

Solve the system for x and y.

$2x + y = 4$

$2x^2 + 3xy - y^2 = -4$

Solution

We solve the linear equation for y and obtain,

$$y = -2x + 4 \tag{2}$$

Substitution of (2) into the second equation yields

$$2x^2 + 3x(-2x + 4) - (-2x + 4)^2 = -4$$

Simplifying, we have

$$8x^2 - 28x + 12 = 0$$

factoring yields

$$4(2x - 1)(x - 3) = 0$$

Thus, we have two solutions for x, namely $x = \frac{1}{2}$ or $x = 3$. To find y we substitute these values of x into (2). Thus, when $x = \frac{1}{2}$, we have $y = -2(\frac{1}{2}) + 4 = 3$, and when $x = 3$, we have $y = -2(3) + 4 = -2$. Thus, our two solutions are $(\frac{1}{2}, 3)$ and $(3, -2)$. We leave it to you to check these solutions. The graph is given in Figure 3.

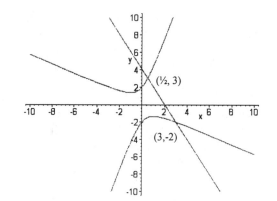

Figure 3: The Graphs of $2x + y = 4$ and $2x^2 + 3xy - y^2 = -4$

The graph of $2x^2 + 3xy - y^2 = -4$ consists of two curve portions shown in Figure 3, and is called a *hyperbola*.
■■■

Example 3
Solve the following system for x and y.
$$-x^2 + 2y^2 = 46$$
$$xy = -10$$

Solution
Neither of the given equations is linear. However, it is easy to solve for one of the unknowns in the second equation, suppose we choose to solve for x (you could just as well solve for y), then we obtain

$$x = -10/y \qquad\qquad (3)$$

We substitute using (3) for x in the first equation and obtain,

$$-(-10/y)^2 + 2y^2 = 46$$

or

$$-100/y^2 + 2y^2 = 46$$

We clear fractions by multiplying by the *LCD* which is y^2 and obtain

$$-100 + 2y^4 = 46y^2$$

or rewriting as

$$2y^4 - 46y^2 - 100 = 0$$

or equivalently as

$$y^4 - 23y^2 - 50 = 0$$

We factor this into

$$(y^2 - 25)(y^2 + 2) = 0$$

Therefore, the first factor yields

$$y^2 - 25 = 0$$

or $y = -5$ or $y = +5$.

The second factor yields the imaginary solutions $y = \pm i\sqrt{2}$. (Verify this!) Since we are only interested in real solutions (points where the curves intersect), we omit consideration of these.

When $y = -5$, we substitute into (3) and obtain $x = -10/-5 = 2$, and when $y = 5$, we obtain $x = -10/5 = -2$. Thus, the two real solutions are $(-2, 5)$ and $(2, -5)$. (We leave it as an exercise for you to show that the two non-real solutions are $(-5i\sqrt{2}, -i\sqrt{2})$ and $(5i\sqrt{2}, i\sqrt{2})$.). The graph illustrating the curves and their intersections is given in Figure 4.

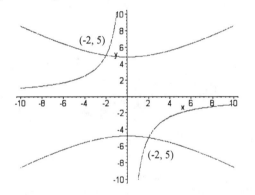

Figure 4: The Graphs of $-x^2 + 2y^2 = 46$ and $xy = -10$

(Note, the graph of $xy = -10$ is also called a hyperbola, it consists of two curved pieces in quadrants II and IV, as shown in Figure 4.)
■■■

Sometimes, there are alternative ways to solve a system of equations. In the next example, we illustrate three different methods.

Example 4
Solve the following system for x and y.
$3x + 2y = 18$
$xy = 12$

Solution

Method 1

Here we follow the general advice that we solve the linear equation for one of the unknowns and then substitute into the non-linear equation. Solving the first equation for y yields

$$y = \frac{-3x + 18}{2} \qquad (4)$$

we next substitute into the second equation to obtain

$$x\left(\frac{-3x + 18}{2}\right) = 12$$

Clearing fractions and rewriting, we obtain

$$3x^2 - 18x + 24 = 0$$

This factors into

$$3(x - 4)(x - 2) = 0$$

Thus, $x = 2$ or $x = 4$. We substitute each of these values for x into (4) to find the corresponding y values. When $x = 2$, $y = (-3(2) + 18)/2 = 6$, and when $x = 4$, $y = (-3(4) + 18)/2 = 3$. Thus, our solutions are (2,6) and (4,3). The graphs is given in Figure 5.

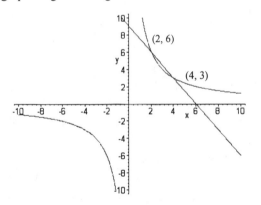

Figure 5: The Graphs of $3x + 2y = 18$ and $xy = 12$

Method 2

Solve the second equation for y yielding

$$y = 12/x$$

We next substitute for y in the first equation

$$3x + 2(12/x) = 18$$

Clearing fractions and rewriting yields

$$3x^2 - 18x + 24 = 0$$

The rest of the solution continues as in Method 1.

Method 3
Multiply the first equation by x yielding

$$3x^2 + 2xy = 18x$$

Since $xy = 12$,

We have

$$3x^2 + 2(12) = 18x$$

or

$$3x^2 - 18x + 24 = 0$$

The rest of the solution continues as in Method 1.

■■■

Method of Elimination

In the examples illustrated above, substitution was used in one form or another. However, there are a class of problems for which the method of elimination is indeed effective. In particular, if each equation of the system has only terms in x^2 and y^2, as the next example illustrates.

Example 5
Solve the following system for x and y.
$$3x^2 + 4y^2 = 19$$
$$-2x^2 + 5y^2 = 18$$

Solution
We observe that each equation has only terms in x^2 and y^2, therefore we use the method of elimination. We choose to eliminate x^2 (you could just have well chosen y^2). We multiple each term in the first equation by 2 and each term in the second equation by 3, yielding,

$$6x^2 + 8y^2 = 38$$
$$-6x^2 + 15y^2 = 54$$

Adding the two equations gives

$$23y^2 = 92$$

or

$$y^2 = 4$$

or

$$y = \pm 2$$

When $y = -2$ we substitute into any of the above equations, say the first equation, to obtain

$$3x^2 + 4(-2)^2 = 19$$

or
$$3x^2 = 3$$

or
$$x^2 = 1$$

or
$$x = \pm 1$$

Thus when $y = -2$, we obtain the two solutions $(-1,-2)$ and $(1, -2)$.

Similarly, when $y = 2$, we substitute and obtain,

$$3x^2 + 4(2)^2 = 19$$

or
$$3x^2 = 3$$

or
$$x^2 = 1$$

or
$$x = \pm 1$$

Thus when $y = 2$, we obtain two more solutions $(-1,2)$ and $(1, 2)$. Therefore, we have four solutions in all, $(-1,-2)$, $(1, -2)$, $(-1,2)$, and $(1, 2)$. The graph is indicated in Figure 6.

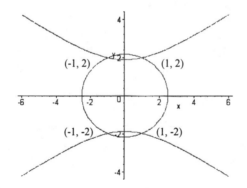

Figure 6: The Graphs of $3x^2 + 4y^2 = 19$ and $-2x^2 + 5y^2 = 18$

■■■
Example 6
Solve the following system for x and y.
$y = \frac{1}{2}x^2 - 3$
$x^2 + y^2 = 9$

Solution
We use the first equation and substitute into the second equation for y to obtain

$$x^2 + (\tfrac{1}{2}x^2 - 3)^2 = 9$$

Multiplying out and combining like terms yields,

$$\frac{1}{4}x^4 - 2x^2 = 0$$

Clearing fractions, we have

$$x^4 - 8x^2 = 0$$

factoring, we have

$$x^2(x^2 - 8) = 0$$

Whose solutions are $x = 0$ or $x = \pm\sqrt{8} = \pm 2\sqrt{2}$. Substituting into the first equation we obtain the corresponding y values. When $x = 0$, we have $y = -3$, and when $x = \pm 2\sqrt{2}$, we find $y = 1$. The three solutions are $(0, -3)$, $(-2\sqrt{2}, 1)$ and $(2\sqrt{2}, 1)$. The graph is given in Figure 7.

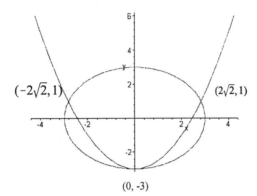

Figure 7: The Graphs of $y = \frac{1}{2}x^2 - 3$ and $x^2 + y^2 = 9$

∎∎∎

Example 7
Find the real solutions of the following system
$x^2 + y^2 = 9$
$x - y = 5$

Solution
Solving the second equation for x gives

$$x = y + 5$$

Substitution into the second equation gives

$$(y + 5)^2 + y^2 = 9$$

or

$$2y^2 + 10y + 16 = 0$$

or

$$y^2 + 5y + 8 = 0$$

190

The solutions to this quadratic equation are $y = \frac{-5 \pm i\sqrt{7}}{2}$. (Verify this!) These are complex numbers, therefore there are no real solutions. From $x = y + 5$, we find the corresponding x-values are $x = \frac{5 \pm i\sqrt{7}}{2}$, and the complex solutions are $\left(\frac{5+i\sqrt{7}}{2}, \frac{-5+i\sqrt{7}}{2} \right)$ and $\left(\frac{5-i\sqrt{7}}{2}, \frac{-5-i\sqrt{7}}{2} \right)$. As mentioned above, if we are only interested in real solutions, i.e., the intersections of the graphs, we do not have to find the complex solutions. The graph is given in Figure 8, notice that the curves have no points of intersection - no real solutions!

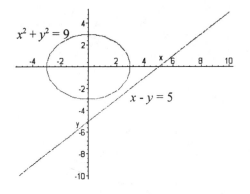

Figure 8: The Graphs of $x^2 + y^2 = 9$ and $x - y = 5$

■■■

Example 8
Solve the system
$x^2 + 3xy + y^2 = 11$
$x^2 + y^2 = 5$

Solution
Notice that we can rewrite the first equation as

$$x^2 + y^2 + 3xy = 11$$

and from the second equation, we have that $x^2 + y^2 = 5$, therefore, we can substitute for the first two terms, and we have the equation

$$5 + 3xy = 11$$

or

$$xy = 2$$

(Observe that if we subtracted the second equation from the first, we would also have obtained this equation.)

Therefore, we replace the first equation with this equation and solve the equivalent system

$$x^2 + y^2 = 5$$
$$xy = 2$$

solving this second equation for y yields $y = 2/x$, and substitution into the first of these equations yields

$$x^2 + (2/x)^2 = 5$$

or clearing fractions and rewriting, we have

$$x^4 - 5x^2 + 4 = 0$$

This may be factored as

$$(x^2 - 4)(x^2 - 1) = 0$$

Thus, we find that $x = -1, 1, -2, 2$, and the corresponding y values found from $y = 2/x$ are $y = -2, 2, -1, 1$, and the solutions are $(-1,-2)$, $(1,2)$, $(-2,-1)$ and $(2,1)$. The graph illustrating these equations is given in Figure 9.

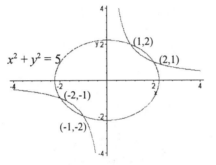

Figure 9: The Graphs of $x^2 + 3xy + y^2 = 11$ and $x^2 + y^2 = 5$

■■■

The above problem was "rigged" in the sense that the second equation is also part of the first equation, so the substitution of the second into the first gives considerable simplification. If any of the coefficients were different, this method would not work.

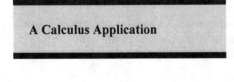

A Calculus Application

An important problem examined in calculus is the optimization of a function, that is, locating the points at which a function is either maximized or minimized. In determining such an optimal solution it is often necessary to solve a system of equations which may or may not be linear. In the next few examples, we illustrate how we may solve such systems. We remark, that in addition to the usual variable names, it is customary in the calculus to use the Greek letter λ (lambda) for one of them, which is known as the Lagrange multiplier.

There are many ways to solve these systems of equations, we give one procedure which works on most of the systems you will encounter in the calculus.

Example 9

Solve the following system of equations for x, y and λ.

$y - 5\lambda = 0$

$x - 2\lambda = 0$

$5x + 2y = 1200$

Solution

This problem is a linear system in the three unknowns. There are many efficient methods to solve such systems. However, our approach is one that will work nicely in the non-linear examples that follow.

If we solve the first and second equations for λ, we have

$$y/5 = \lambda$$

and

$$x/2 = \lambda$$

Since each equation is equal to λ, they are equal to each other so we have

$$y/5 = x/2$$

or

$$y = \frac{5}{2}x$$

Substituting for y in the third equation, we have

$$5x + 2(\frac{5}{2}x) = 1200$$

$$10\,x = 1200$$

$$x = 12$$

since $y = \frac{5}{2}x$, we have

$$y = 5/2(12) = 30$$

and $\lambda = x/2 = 12/2 = 6$.

■■■

In the preceding example, our strategy was to first solve the equations containing λ for that variable. When this was done, we were able to solve for y in terms of x and then substitute into the equation not containing λ to obtain x and then substitute to obtain y and λ. We shall use a slight generalization of this strategy to solve similar non-linear systems.

We shall only be interested in solutions whose variables are positive. This follows from the applications in which these problem arise where the variables represent a resource or measure which must be positive.

Example 10
Find the solutions to the following system for x, y, z and λ which are positive.

$yz = 2\lambda$
$xz = 4\lambda$
$xy = 5\lambda$
$2x + 4y + 5z = 60$

Solution
We solve the first three equations for λ to obtain

$$yz/2 = \lambda \qquad \qquad \qquad ❶$$

$$xz/4 = \lambda \qquad \qquad \qquad ❷$$

$$xy/5 = \lambda \qquad \qquad \qquad ❸$$

Setting ❶ = ❷ we have

$$yz/2 = xz/4$$

clearing fractions,

$$2yz = xz$$

or

$$2yz - xz = 0$$

or

$$z(2y - x) = 0$$

since all the variables must be positive, we have

$$2y = x$$

or

$$y = x/2$$

Now we set ❶ = ❸ we have

$$yz/2 = xy/5$$

clearing fractions, we have

$$5yz = 2xy$$

or

$$5yz - 2xy = 0$$

or

$$y(5z - 2x) = 0$$

or

$$5z - 2x = 0$$

or

$$z = 2x/5$$

Note that we have both y and z in terms of x. We now substitute for y and z in the fourth of the original equations, the only equation not containing the variable λ.

$$2x + 4y + 5z = 2x + 4(x/2) + 5(2x/5) = 60$$

or

$$6x = 60$$

thus,

$$x = 10$$

$$y = x/2 = 5$$

$$z = 2x/5 = 4$$

and

$$\lambda = yz/2 = 10.$$

■■■

We summarize the procedure.

1. Solve each of the equations containing λ for that variable.

2. Set these equations equal to each other to obtain all variables in terms of one of them, in the above example, y and z in terms of x. In particular, we can set equations ❶ = ❷ and equations ❶ = ❸. This will determine y and z in terms of x

3. Substitute into the equation not containing λ and solve for the unknown, and then substitute back to find the remaining unknowns.

Example 11
Find the solutions to the following system for x, y, z and λ which are positive.
$y + 2z - \lambda yz = 0$
$x + 2z - \lambda xz = 0$
$2y + 2x - \lambda xy = 0$
$xyz = 32$

Solution
We solve each of the first three equations for λ. The first equation becomes
$$y + 2z = \lambda yz$$
or

$$\frac{y + 2z}{yz} = \lambda \qquad\qquad ❶$$

similarly, the other two equations become

$$\frac{x + 2z}{xz} = \lambda \qquad\qquad ❷$$

and

$$\frac{2y + 2x}{xy} = \lambda \qquad\qquad ❸$$

Setting equation ❶ = ❷, we have

$$\frac{y+2z}{yz} = \frac{x+2z}{xz}$$

Clearing fractions we have

$$xz(y + 2z) = yz(x + 2z)$$

or

$$xyz + 2xz^2 = xyz + 2yz^2$$

or

$$2xz^2 - 2yz^2 = 0$$

or

$$2z^2(x - y) = 0$$

thus, (the solution $z = 0$ is not considered since we are looking for positive solutions)

$$y = x$$

We next set equations ❶ = ❸, we have

$$\frac{y+2z}{yz} = \frac{2x+2y}{xy}$$

clearing fractions,

$$xy(y + 2z) = yz(2x + 2y)$$

or

$$xy^2 + 2xyz = 2xyz + 2y^2z$$

or

$$xy^2 - 2y^2z = 0$$

or

$$y^2(x - 2z) = 0$$

or

$$z = x/2$$

Thus, we now have both y and z in terms of x so we substitute into the fourth of the original equations, the one that does not contain λ. We have

$$xyz = x(x)(x/2) = 32$$

$$x^3 = 64$$

or

$$x = 4$$

$$y = x = 4$$

$$z = x/2 = 2$$

and

$$\lambda = \frac{y+2z}{yz} = \frac{4+2(2)}{4\cdot 2} = 1$$

■■■

Exercise set 18

In exercises 1 - 3 solve the given system of equations for *the real values of x* and *y*.

1. $y = 3x - 3$
   $xy = 6$

2. $2x + y = 2$
   $xy = -4$

3. $x - 2y = 0$
   $xy = 4$

4. $2x - y = 0$
   $xy = 6$

5. $2x - y = 7$
   $3x^2 - 2xy + y^2 = 34$

6. $3x + y = 9$
   $x^2 - 2xy + 3y^2 = 19$

7. $x - 4y = -11$
   $3x^2 + 2x - 3xy + 2y^2 = 14$

8. $3x - 4y = 17$
   $3x^2 + 2xy - y^2 + 2y = 7$

9. $2x + 5y = 4$
   $2x^2 - 5x - 3xy + 3y^2 = 63$

10. $2x^2 + y^2 = 6$
    $3x^2 - y^2 = 4$

11. $3x^2 - y^2 = 28$
    $2x^2 + y^2 = 32$

12. $4x^2 - 2y^2 = 18$
    $2x^2 + 3y^2 = 117$

13. $2x^2 - 3y^2 = 6$
    $3x^2 + 5y^2 = 47$

14. $5x^2 - 2y^2 = 62$
    $3x^2 + 5y^2 = 93$

15. $2x^2 + 3xy + y^2 = 35$
    $xy = 6$

16. $3x^2 - 2xy + 4y^2 = 3$
    $2xy = 1$

17. $4x^2 + 10xy + 25y^2 = 76$
    $5xy = 12$

18. $50x^2 + 10xy - 3y^2 = -6$
    $5xy = 2$

19. $y - 3x = 2$
    $y = 2x^2 - 3x + 6$

20. $x + 2y = 13$
    $x = 2y^2 - 3y - 32$

21. $x^2 + y^2 = 25$
    $y = x^2 - 5$

22. $x^2 + y^2 = 169$
    $y = x^2 - 13$

23. $x^2 + y^2 = 25$
    $x^2 + 2x + y^2 - 5y = 11$

24. $x^2 + y^2 = 10$
    $x^2 - 5x + y^2 + 7y = 2$

25. $x^2 - 4xy + 2y^2 = -2$
    $x^2 + 2y^2 = 6$

26. $3x^2 + 2xy + 5y^2 = 38$
    $3x^2 + 5y^2 = 32$

27. $2x^2 + 5xy - 3y^2 = 60$
    $2x^2 - 3y^2 = 20$

28. $5x^2 + 2xy + 3y^2 = 35$
    $5x^2 + 3y^2 = 47$

In exercises 29 - 36 find the real solutions to the given system of equations for which *all* the variables are *positive*

29. $5x - 3\lambda = 0$
    $5y - 2\lambda = 0$
    $2x + 3y = 120$

30. $y - 6\lambda = 0$
    $x - 2\lambda = 0$
    $6x + 2y = 180$

31. $yz - 10\lambda = 0$
    $xz - 20\lambda = 0$
    $xy - 25\lambda = 0$
    $10x + 20y + 25z = 240$

32. $yz - 6\lambda = 0$
    $xz - 2\lambda = 0$
    $xy - 10\lambda = 0$
    $5x + 10y + 25z = 500$

33. $8x - 10\lambda = 0$
    $4y - 20\lambda = 0$
    $8z - 50\lambda = 0$
    $10x + 20y + 50z = 680$

34. $xy^2z^2 - 3x\lambda = 0$
    $x^2yz^2 - y\lambda = 0$
    $x^2y^2z - 3z\lambda = 0$
    $3x^2 + y^2 + 3z^2 = 36$

35. $2y + 2z - \lambda yz = 0$
    $2x + z - \lambda xz = 0$
    $y + 2x - \lambda xy = 0$
    $xyz = 4$

36. $2xy^3z^4 - \lambda = 0$
    $3x^2y^2z^4 - 2y\lambda = 0$
    $4x^2y^3z^3 - 3z^2\lambda = 0$
    $x + y^2 + z^3 = 29$

37. Sketch the graph of a line and a circle showing (a) no point of intersection, (b) one point of intersection, (c) two points of intersection.

38. Sketch the graph of a parabola and a line, showing all possibilities regarding the number of points of intersection.

39.   Sketch the graph of a parabola and a circle, showing all possibilities regarding the number of points of intersection.

40.   Given the system of equations $ax^2 + by^2 = c$ and $dx^2 + ey^2 = f$. Solve this system of equations for x and y. Under what conditions will the solutions be real?

POSTTEST 18- Time 20 minutes

Each question is worth two points. Solve the given system of equations for the real values of x and y.

1. $3x + 2y = 1$
   $8x^2 - 4xy + 2y^2 = 14$

2. $3x^2 + 2y^2 = 20$
   $xy = 4$

3. $5x^2 - 3y^2 = 33$
   $3x^2 + 4y^2 = 43$

4. $y = 5x^2 - 3$
   $y^2 - 2x^2 = 281$

5. $2x^2 + 4xy + 3y^2 = 6$
   $2x^2 + 3y^2 = 14$

---

19. APPENDIX - Arithmetic of Signed Numbers

> » **Absolute Value**
> » **Addition of Signed Numbers**
> » **Subtraction of Signed Numbers**
> » **Multiplication of Signed Numbers**
> » **Division of Signed Numbers**

PRETEST 19- Time 5 minutes

Each question is worth two points. Simplify the given expression:

1. 10 - 21

2. (-23) + (-17)

3. (-17) + (85)

4. 15 - (-8)

5. -18 - (-22)

6. 21 - 35

7. -3(-4)(-2)

8. -2{3- (-2)(12 - 15)}

9. -(-2)(-3)(-5)

10. $\dfrac{-4(12)}{-3(-2)}$

In this section we quickly review the addition, subtraction, multiplication and division of signed numbers. Let us recall that negative numbers have many important applications. Certainly with respect to finance, a negative number may represent a loss. Suppose, for example, a company reported its quarterly profits over a one year period (in millions of dollars) as 168, 54, -28 and 14. Notice the -28 reported for the third quarter represents a loss of 28 million dollars. What is the company's profit for the entire year? Undoubtedly, you computed it to be 208 million dollars. You probably had no difficulty absorbing the loss of 28 in the sum. Let us now formulate the rules for the arithmetic operations.

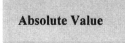
Absolute Value

Recall that the absolute value of a number is the magnitude of a number. For example the absolute value of -5, written |-5| = 5, |17| = 17, |-22| = 22, |0| = 0. In fact, if you recall the square root symbol, we could define the absolute value of any number x as $|x| = \sqrt{x^2}$.

We are now ready to define the arithmetic of signed numbers. (Remember, if there is no sign in front of a number the number is positive, for example, 2 = + 2.)

<table>
<tr><td>**Addition of Signed Numbers**</td><td>Note: We use the word *add* in the sense it was used with positive numbers, for example, 9 + 7 = 16.</td></tr>
</table>

Rule 1: To add two numbers of the same sign add their absolute values and attach their common sign to the sum obtained.

Example 1

Simplify (a) 12 + 19 (b) (-15) + (-28).

Solution

(a) since both numbers are positive their sum is 31.

(b) since both numbers are negative we sum their absolute values. $|-15| = 15$, $|-28| = 28$, so 15 + 28 = 43. We now attach the original sign to obtain (-15) + (-28)= -43.

■■■

Note that (-15) +(-28) = -43 makes perfectly good sense. Suppose you owed 15 dollars and then borrowed an additional 28 dollars, your total debt is then 43 dollars, debt being represented by a negative number.

Rule 2: To add two numbers with opposite signs, compute their absolute values and subtract the smaller absolute value from the larger, attaching the sign of the larger number to this difference.

Note: We used *subtract* in the sense you were taught with positive numbers , we subtract a smaller positive number from a larger one for example, 12 - 9 = 3.

Example 2

Simplify (a) 16 + (-9)   (b) -16 + 9.

Solution

(a) $|16| = 16$, $|-9| = 9$, thus, 16 -9 = 7 and since the sign of the number with the larger absolute value is positive, we have 16 + (-9) = 7.

(b)$|-16| = 16$, $|9| = 9$, thus, 16 - 9 = 7, and since the sign of the number with the larger absolute value is negative, we have -16 + 9 = -7.

■■■

What if the two numbers have opposite sign and the same absolute value, for example 6 + (-6)? Suppose you have $6 and then lose $6, what remains? Clearly the sum of two equal and opposite numbers is 0.

The rules given above are for the sum of two numbers, what if you are asked to sum more than two numbers? All we need do is sum them two at a time as illustrated in the next example.

Example 3

Simplify 3 + (-12) + (-9) + 28.

Solution

$$3 + (-12) + (-9) + 18 = -9 + (-9) + 28 = -18 + 28 = 10.$$

Alternately, we could rewrite the expression by grouping the positive numbers together and the negative number together,

$$3 + (-12) + (-9) + 28 = 3 + 28 + (-12) + (-9) = 31 + (-21) = 10.$$

(The justification for rewriting the numbers is given in Section 1, where we review the properties of real numbers.)
■■■

With addition of signed numbers defined, we are now able to define the rule for the subtraction of *signed* numbers.

Subtraction of Signed Numbers

Rule 3: Let a and b be any real numbers, then we define $a - b = a + (-b)$.

This means that to subtract b from a, (1) change the subtraction operation to addition, (2) change the sign of b and, (3) then use the rules of addition given in Rules 1 and 2.

Example 4
Simplify each of the following. (a) 16 - (-5) (b) 5 - 16 (c) -21 - (-23).

Solution
(a) $16 - (-5) = 16 + (5) = 21$
(b) $5 - 16 = 5 + (-16) = -11$
(c) $-21 - (-23) = -21 + (23) = 2$
■■■

Note, that the expression 12 - 5 can be interpreted as the subtraction of 5 from 12 or the addition of 12 and -5, that is, $12 - 5 = 12 + (-5)$, they both yield the same result, 7.

How should multiplication of signed numbers be defined? Consider the following products:

$$5(4) = 20$$
$$5(3) = 15$$
$$5(2) = 10$$
$$5(1) = 5$$
$$5(0) = 0$$
$$5(-1) = ?$$

Multiplication of Signed Numbers

Notice as we reduced the second factor, the product became smaller. Thus, when the second factor is a negative number it only makes sense for the product to be negative since it needs to be smaller then 0. Alternately, if we think of multiplication as repeated addition $5(-1) = (-1) + (-1) + (-1) + (-1) + (-1) = -5$. This gives the following rule.

Rule 4: To multiply a positive number by a negative number, multiply their absolute values and attach a negative sign to this product.

Thus $(5)(-4) = -20$, $-10(42) - -420$, and so on.

What about the product of two negative numbers. Let us follow the approach used above. Consider

$$(-5)(4) = -20$$
$$(-5)(3) = -15$$
$$(-5)(2) = -10$$
$$(-5)(1) = -5$$
$$(-5)(0) = 0$$
$$(-5)(-1) = ?$$

Look at the pattern, as the second factor becomes smaller the product becomes larger thus after $(-5)(0) = 0$, the product must continue to increase and therefore be positive. This gives the next rule.

Rule 5: The product of two negative numbers is the product of their absolute values.

Thus, the product of two negative numbers is positive.

Example 5
Simplify
(a) $(-2)(3)$ (b) $(-2)(-3)$ (c) $(-2)(-3)(-4)$ (d) $(-2)(-3)(-4)(-5)$

Solution
(a) $(-2)(3) = -6$
(b) $(-2)(-3) = 6$
(c) $(-2)(-3)(-4) = 6(-4) = -24$
(d) $(-2)(-3)(-4)(-5) = 6(-4)(-5) = -24(-5) = 120$.
■■■

Note that the sign of a product of an odd number of negative numbers is negative and the product of an even number of negative numbers is positive (why?).

Two other important observations are $-b = -1(b)$ and $b = 1(b)$ These appear trivial, but have useful applications. We restate this observation as a rule.

Rule 6: If a negative sign is in front of a parenthesis, the parenthesis may be removed provided the sign of every term within the parenthesis is multiplied by -1. If a positive sign is in front of a parenthesis the parenthesis may be removed leaving the terms within the parenthesis as they were..

Example 5
Remove the parenthesis from the given expression and then simplify.
(a) $-(4 - 7)$ (b) $(17 - 21)$

Solution
(a) -(4 - 7) = -4 + 7 = 3   (b) (17 - 21) = 17 - 21 = -4.
■■■

| **Division of Signed Numbers** | The rules for division follow from the rules for multiplication, since for $b \neq 0$, $a/b = c$ means $a = bc$. Therefore, we see that if a and b are both positive or both negative c must be positive, and if a and b have opposite signs c must be negative. For example $-12/4 = c$ means $4c = -12$. Clearly c must be -3. We summarize in the next rule. |

Rule 7 : If the numerator and denominator of a fraction have the same sign, the sign of their quotient is positive, if they have different sign, their quotient is negative.

Example 6
Simplify (a) -24/8   (b) 15/(-3)   (c) - (-25/-5) (d) (-3)(-4){(-12)/(2)}

Solution
(a) -24/8 = -3 (b) 15/(-3) = -5   (c) -(-25/-5)= -(5) = -5 (d) (-3)(-4){(-12)/(2)} = 12{-6} - -72
■■■

Exercise set 19
In exercises 1 - 25 simplify the given expression.

1.   -6 + (-11)

2.   -22 + 17

3.   19 + (-43)

4.   23 - (-11)

5.   14 + (-32) +(35) + (-3) + (-15)

6.   42 - 63

7.   -(21) - (-21)

8.   -47 - 12

9.   (-17) - (-32) - (11) - (48)

10.   21 - (-87) + (-32) - (-67) + 25

11.   -23 - (-19) + (28) + (-7) - (-51)

12.   (-2)(-5)

13.   (14)(-3)
14.   -2(-3)(-5)(-6)

15.   (-2)(-5)(10)(-3)

16.   (-5)(-4) - 2(-15)

17.   -(-4)(-5) + 6(-8)

18.   12(-5) - (-6)(10)

19.   25/(-5)

20.   (-15)/(-3)

21.   -(-42)/(-7)

22.   -3(-18)/(-9)

23.   12(-6)/(-3)

24.   -4(-8)/2 + (-3)(4) - (-5)

25.   12(-4)(-3)/(-36) + 10(-8) - (-7)(-9)

In exercises 26 - 30, compute the result two ways (1) by first simplifying within the parenthesis and (2) remove the parenthesis and then simplifying.

26.   -(12 - 23)

27.   -(32 - 98)

28.   - (14 - 28)

29.   -(74 - (-5))

30.   -(14 - 11 + 2(8))

POSTTEST 20- Time 5 minutes

Each question is worth one point. Simplify the given expression:

1. -12 + (-15)

2. (33) + (-14)

3. 81 - 78

4. -12 - (-32)

5. 41 -32 + 17 - (-11)

6. -3(-2)(-5)(4)

7. $\dfrac{14(-6)}{-7(-3)}$

8. -3{16-2 -3[4-(-12)]}

9. {(-2)(-8)}/{-4+2(-5)}

10. $-\dfrac{(-12)(-4)}{(-6)(-8)}$

20. ANSWERS TO PRE AND POST TESTS

Pretest 1
1. (c) 2. 18 3. 18 4. -16 5. 222. 6. 38 7. $7x^2y^7$ 8. Associative 9. Distributive 10. Commutative

Posttest 1
1. (a) & (c) 2. 21 3. 48 4. -16 5. 135 6. 323 7. $7x^9y^3$ 8. Associative 9. Distributive 10. Commutative

Pretest 2
1. 8/27 2. 1/8 3. x^9 4. a^{12} 5. 9 6. x^9 7. 1 8. 9/8 9. $\dfrac{a^8c^2}{b^{10}}$ 10. $\dfrac{x^{15}}{8x^{21}}$

Posttest 2
1. 81/16 2. 1/16 3. a^8 4. b^{15} 5. 16 6. a^8 7. 1 8. $\dfrac{y^3}{x^2}$ 9. $\dfrac{y^{15}z^3}{x^{21}}$ 10. $\dfrac{x^{12}}{4y^{14}}$

Pretest 3
1. 3 2. x^6 3. $6\sqrt{2}$ 4. $8\sqrt[3]{2}$ 5. $2x^4y^5\sqrt{6xy}$ 6. 5 7. $6\sqrt{2}$ 8. $13\sqrt{2}$ 9. $6\sqrt{3}$ 10. $4\sqrt{5}$

Posttest 3
1. $2\sqrt[4]{2}$ 2. y^6 3. $9\sqrt{3}$ 4. $10\sqrt[4]{2}$ 5. $3x^4y^3\sqrt[3]{2y}$ 6. 5 7. $9\sqrt{5}-6\sqrt{6}$ 8. $11\sqrt{2}$ 9. 8 10. $9\sqrt{2}$

Pretest 4.
1. $16-18\sqrt{5}$ 2. $6\sqrt{3}-12\sqrt{2}$ 3. $8+7\sqrt{2}$ 4. $30-12\sqrt{6}$ 5. 33 6. $(x-2\sqrt{3})(x+2\sqrt{3})$ 7. $3\sqrt{2}$
8. $\dfrac{\sqrt{3}}{3}$ 9. $5\sqrt[3]{2}$ 10. $10+4\sqrt{3}$

Posttest 4
1. $21-9\sqrt{3}$ 2. $20\sqrt{2}-30\sqrt{5}$ 3. $12+9\sqrt{5}$ 4. $141-24\sqrt{30}$ 5. -22 6. $(x+2\sqrt{5})(x-2\sqrt{5})$ 7. $\dfrac{8\sqrt{5}}{5}$
8. 60 9. $6\sqrt[3]{3}$ 10. $6\sqrt{5}+4\sqrt{7}$

Pretest 5
1. -15 2. $2\sqrt{6}i$ 3. $3+10i$ 4. $21+i$ 5. $5-12i$ 6. 74 7. $-5i/3$ 8. $10+4i$ 9. $\pm 2i$ 10. $\dfrac{\pm 5\sqrt{2}}{6}i$

Posttest 5
1. -28 2. $3\sqrt{5}i$ 3. $1+i$ 4. $47+i$ 5. $7-24i$ 6. 40 7. $\dfrac{-12}{5}i$ 8. $12-9i$ 9. $\pm 3i$ 10. $\dfrac{\pm 2\sqrt{10}}{15}i$

Pretest 6
1. 3 2. 1/4 3. 81 4. 1/8 5. 27 6. 16 7. 2 8. $\sqrt[6]{32}$ 9. \$1485.95 10. 8.01%

Posttest 6
1. 5 2. ½ 3. 125 4. 1/81 5. 1/16 6. 4 7. 10 8. $\sqrt[6]{243}$ 9. \$1819.40 10. 11.89%

Pretest 7

1. $8x^2y^3(2 + 3y)$ 2. $6x^{1/2}(2x + 3)$ 3. $-3x^3(x^2 + 3)(x^2 - 1)^{14}$ 4. $\frac{x(3x^2 - 2)}{\sqrt{x^2 - 1}}$ 5. $(4x - 5)(3x - 4)$

6. $(4xy - 5)(5xy + 4)$ 7. $-5(2x - 1)(2x - 3)$ 8. $(2x - 5)(2x + 5)$ 9. $(3xy - 7z)(3xy + 7z)$ 10. $(3a - 2)(x + 4)$

Posttest 7

1. $5x^3y^2(5y^2 + 6x)$ 2. $5x^{\frac{1}{4}}(3 + 4x)$ 3. $9x^2(x^2 + 3)^{11}(5x^2 + 6)$ 4. $\frac{3x}{\sqrt{2x^2 + 1}}$ 5. $(4x - 3)(6x + 5)$

6. $(6x - 5y)(3x - 4y)$ 7. $-(6x - 5)(4x - 3)$ 8. $(3x - 7)(3x + 7)$ 9. $(5x - 9yz)(5x + 9yz)$

10. $2(2a - 1)(x - 2)(x + 2)$

Pretest 8

1. 1   2. -3   3. 3   4. 7   5. -2/3   6. 59/158 ≈0.37342   7. 20   8. 45   9. 4/3   10. $(2x - 5)/3$

Posttest 8

1. 1   2. 2   3. -3   4. 15   5. 17/2   6. -32/7   7. 2/9   8. -270/109 ≈-2.47706   9. $(3x - 9)/4$ 10. 6 ft

Pretest 9

1. 0, 5/4   2. 0, 4   3. ±7/3   4. $\frac{\pm\sqrt{3}}{2}$ 5. 2, -13/6   6. $\pm 2\sqrt{2}i$ 7. $\frac{\pm 2\sqrt{2}}{3}$ 8. $5 \pm 3\sqrt{2}$ 9. $5 \pm 3\sqrt{3}i$ 10. $\frac{5 \pm 2\sqrt{5}}{2}$

Posttest 9

1. 0, 3/7   2. 0, 9/2   3. ±9/4   4. ±3/2   5. ±1   6. $\pm 2\sqrt{5}i$ 7. $\frac{\pm\sqrt{6}}{4}$ 8. $-3 \pm 2\sqrt{6}$ 9. $5 \pm 2\sqrt{3}i$ 10. $\frac{-4 \pm 2\sqrt{3}}{5}$

Pretest 10

1. 16   2. 9/4   3. $5 \pm 2\sqrt{7}$ 4. $\frac{-5 \pm 3\sqrt{5}}{2}$ 5. $\frac{6 \pm 4\sqrt{3}}{3}$

Posttest 10

1. 36   2. 25/4   3. $6 \pm 2\sqrt{10}$ 4. $\frac{-3 \pm 3\sqrt{5}}{2}$ 5. $2 \pm \sqrt{6}$

Pretest 11

1. $\frac{2 \pm \sqrt{14}}{2}$ 2. $2 \pm \sqrt{5}$ 3. $\frac{4 \pm \sqrt{34}}{9}$ 4. $\frac{3 \pm \sqrt{3}i}{3}$ 5. 1.664 or 198.336 sec.

Posttest 11

1. $\frac{3 \pm \sqrt{41}}{8}$ 2. $\frac{1 \pm \sqrt{33}}{4}$ 3. -1, 5/4   4. $\frac{5 \pm \sqrt{23}i}{8}$ 5. 7.129 sec.

Pretest 12

1. Two real rational roots   2. Two real conjugate irrational roots   3. $x^2 - x - 6 = 0$ 4. 2/3   5. 5/3

Posttest 12

1. Two complex conjugate roots   2. Two real rational roots   3. $x^2 + x - 30 = 0$ 4. 5/7   5. -3/7

Pretest 13

1. 16   2. 2   3. 5   4. ±3   5. 1

Posttest 13
1. 32   2. 5   3. 3   4. ±2   5. 2, 6

Pretest 14
1. $x \le -4$ or $x \ge 2/3$ 2. $x < 0$ 3. $x \le 3/2$ or $x > 5$ 4. $-11/4 < x < 4/3$ 5. $x < -4$ or $0 \le x \le 1$ or $2 \le x < 4$

Posttest 14
1. $x \le -3/2$ or $x \ge 2/5$ 2. $3/4 < x < 1$ 3. $-3/2 < x \le 5/2$ 4. $-4/5 < x < 3$ 5. $-5/3 < x \le 0$ or $2/3 \le x < 5/3$ or $x \ge 5/2$

Pretest 15
1. $y = 2x + 5$ 2. $5/2$ 3. $-9/2$ 4. $9/5$ 5. $y = 3x - 11$ 6. $x = 5$ 7. $y = 2$ 8. $5x - 2y = 16$ 9. $3x - 2y = 8$
10. $2x + 3y = 27$

Posttest 15
1. $y = 3x - 3$ 2. $-2/7$ 3. $12/7$ 4. 6   5. $-3x + 2y = 18$ 6. $y = 6$ 7. $x = 5$ 8. $x + 11y = 47$ 9. $2x + 5y = 5$
10. $5x - 2y = 56$

Pretest 16.
1. $x^2 + y^2 - 4x + 6y - 3 = 0$ 2. $r = 3, C(-2, 3)$ 3. $r = 5, C(5, -3)$

4.   5. Ellipse

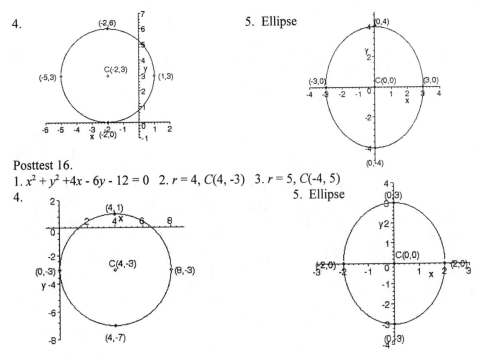

Posttest 16.
1. $x^2 + y^2 + 4x - 6y - 12 = 0$ 2. $r = 4, C(4, -3)$ 3. $r = 5, C(-4, 5)$
4.   5. Ellipse

Pretest 17
1. $(12/23, 14/23)$ 2. $(23/19, -8/19)$ 3. $(36/7, -12/7)$ 4. $(8, -1)$ 5. $(-11/9, 11/31)$

Posttest 17

1. (1, -1)   2. (137/106, 183/106)   3. (-7, -42/5)   4. (-7/41, 115/41)   5. (-63/47, 21/19)

Pretest 18.

1. (-3/13, 97/13), (4, -1)   2. (-4,-3), (4, 3)   3. (-5, -2), (-5, 2) (5, -2), (5, 2)   4. (-1, 7), (1, 7)

5, (-2, -1), (2, 1), $(-\sqrt{2}, -\sqrt{2})$, $(\sqrt{2}, \sqrt{2})$

Posttest 18

1. (-27/37, 59/37), (1, -1)   2. (-2, -2), (2, 2), $(\frac{2}{3}\sqrt{6}, \sqrt{6})$, $(-\frac{2}{3}\sqrt{6}, -\sqrt{6})$ 3. (-3, -2), (-3, 2), (3, -2), (3, 2)

4. (-2, 17), (2, 17)   5. (-1, 2), (1, -2), $(-\sqrt{6}, \frac{\sqrt{6}}{3})$, $(\sqrt{6}, -\frac{\sqrt{6}}{3})$

Pretest 19

1. -11   2. -40   3. 68   4. 23   5. 4   6. -14   7. -24   8. 6   9. 30   10. -8

Posttest 19

1. -27   2. 19   3. 3   4. 20   5. 37   6. -120   7. -4   8. 102   9. -8/7   10. -1

21. ANSWERS TO EXERCISES

Exercise set 1

1. 5   2. -2   3. 6   4. 21 5. 16   6. 8, x, y 7. 2, x, y 8. $x + 2$, $x - 3$ 9. 5, x, $x - 5$, $x + 9$
10. 19, x^3, y, z -11   11. 21   12. -17   13. 36   14. 1   15. 79   16. (a) -27 (b) -27   17. (a) -16 (b) 16
18. 60   19. 406   20. -9/44   21. 12   22. -30   23. -68   24. 0   25. 52   26. Commutative
27. Associative 28. Commutative 29. Distributive 30. Distributive 31. Distributive 32. Identity
33. Identity   34. Inverse   37. $3 + 2n$ 38. $5n -3$ 39. $5n -3$ 40. $3j + 6$ 41. $2(e + 4)$
42. $1.23 = 123/100$ 43. $1.3333.... = 1/3$

Exercise set 2

1. 3^5 2. $(1/2)^3$ 3. 5^7 4. $(-5^7) = -5^7$ 5. $1/16$ 6. $1/27$ 7. 4   8. 27   9. x^4 10. y^6 11. a^5 12. 1   13. 1   14. $\dfrac{x^2}{9y^{14}}$

15. $\dfrac{9}{4xy^3}$ 16. $\dfrac{y+x}{y-x}$ 17. b^3 18. 8   19. 1   20. 125   21. 432 22. 8   23. 729   24. 1024   25. 1/64   26. x^{12}

27. $1/x^{10}$ 28. 5184   29. $64\,x^6$ 30. $128\,x^2$ 31. $\dfrac{8}{27y^{21}}$ 32. $1/x^{11}$

Exercise set 3

1. 3   2. -3   3. -12   4. 30   5. 5/4   6. 11/13   7. -7/8   8. 4/25   9. 3   10. 2/3   11. -3   12. 2   13. 3/2   14. 5
15. 4   16. 13   17. 7   18. 21 19. $17+x$ 20. x^2+9 21. 3   22. 4   23. -7   24. x^3+27 25. 8   26. 2   27. 0   28. -10
29. 16.673   30. 0.272   31. 1.944   32. 2.466   33. 3.362   34. 1.447   35. $2\sqrt{6}$ 36. $4\sqrt{3}$ 37. $4\sqrt{2}$ 38. $2\sqrt{3}$
39. $5\sqrt{2}$ 40. $4\sqrt{5}$ 41. $3\sqrt{7}$ 42. $-2\sqrt{30}$ 43. $18\sqrt{2}$ 44. $9\sqrt{5}$ 45. $-15\sqrt{3}$ 46. $8\sqrt{14}$ 47. $-30\sqrt{3}$ 48. $48\sqrt{5}$
49. $-460\sqrt{2}$ 50. $3\sqrt{2}$ 51. $14\sqrt{2}$ 52. $6\sqrt{5}$ 53. $6\sqrt{10}$ 54. $30\sqrt{2}$ 55. 5   56. 10   57. 5/4   58. $4\sqrt{15}$
59. $5\sqrt{5}$ 60. 24   61. $\dfrac{2}{3}\sqrt{5}$ 62. x^2 63. y 64. z^4 65. x^2y^3 66. $2x^4z^5$ 67. $x^4\sqrt{x}$ 68. $z^7\sqrt{z}$ 69. $x^4z^7\sqrt{xz}$
70. $2x^3\sqrt{2}$ 71. $2x^3y^3\sqrt{2y}$ 72. $2x^3y^3\sqrt{3y}$ 73. $5x^2y^2z^5\sqrt{2xz}$ 74. $4x^{12}y^7z^8\sqrt{5xz}$ 75. $2\sqrt[3]{3}$ 76. $2\sqrt[3]{2}$
77. $3\sqrt[3]{2}$ 78. $4\sqrt[3]{2}$ 79. $-4\sqrt[3]{2}$ 80. $3\sqrt[3]{3}$ 81. $2\sqrt[4]{2}$ 82. $3\sqrt[4]{2}$ 83. 6 84. $5\sqrt[3]{2}$ 85. $2\sqrt[4]{5}$ 86. x^2 87. x^3 88. y^3
89. y^4 90. $x^2\sqrt[3]{x}$ 91. $y^3\sqrt[4]{y^2}$ 92. $y^2z^8\sqrt[3]{z}$ 93. $2y^2$ 94. $2y^2\sqrt[3]{2}$ 95. $2y^2\sqrt[3]{2y}$ 96. $2y^2z^4\sqrt[3]{2y}$
97. $2x^3y^3\sqrt[4]{2x^3z^3}$ 98. $3x^4y^5z^3\sqrt[4]{2y^3z^3}$ 99. $2x^2y^4\sqrt[5]{2y^3}$ 100. 25   101. 16   102. 36   103. $4\sqrt{2}$ 104. $8\sqrt{2}$
105. $6\sqrt{5}$ 106. 2   107. 4   108. 11.314   109. 72.11   110. 6.33×18.97 111. 5   112. 13   113. $7\sqrt{2}$, 9.899
114. $\sqrt{89}$, 9.434   115. $\dfrac{5}{4}\sqrt{41}$, 8.004   116. 5/8, 0.625   117. $11\sqrt{5}$ 118. $-2\sqrt{11}$ 119. $36\sqrt{21}$ 120. $3\sqrt[3]{5}+9\sqrt[3]{9}$
121. $11\sqrt[4]{11}$ 122. $4\sqrt[5]{7}$ 123. $-7\sqrt{2}$ 124. $20\sqrt{3}$ 125. $-2\sqrt{2}$ 126. $-24\sqrt{3}$ 127. $7\sqrt{2}$
128. $-149\sqrt{2}$ 129. 0   130. 0   131. $-6\sqrt{5}$ 132. $2\sqrt{6}$ 133. $27\sqrt{7}$ 134. $20\sqrt{10}$ 135. $-\dfrac{11}{12}\sqrt{6}$
136. $\dfrac{17}{10}\sqrt{5}$ 137. $29\sqrt{3}$ 138. $38\sqrt{2}$ 139. $-31\sqrt{6}$ 140. $28\sqrt{2}-42\sqrt{3}$ 141. $25\sqrt{5}+14\sqrt{7}-30\sqrt{3}$
142. $-20\sqrt{7}$ 143. $8\sqrt{5}+53\sqrt{2}$ 144. $27\sqrt[3]{3}$ 145 0   146. $-15\sqrt[4]{2}$ 147. $27\sqrt[3]{3}-\sqrt{3}$ 148. $-17\sqrt{2}-2\sqrt[3]{3}$
149. $18x\sqrt{2}$ 150. $19xy\sqrt{3x}$ 151. $6x^2y\sqrt{5y}$ 152. $23x^4y^3\sqrt{2y^2}$ 153. 0   154. $-x^6y^2\sqrt[4]{2x}$ 155. 13/5
156. 35/13

Exercise set 4

1. $12 + 3\sqrt{2}$ 2. $6 - 14\sqrt{5}$ 3. $4\sqrt{3} - 6\sqrt{5}$ 4. $-20\sqrt{3} + 10\sqrt{7}$ 5. $4\sqrt{15} + 6\sqrt{21}$ 6. $6 - 5\sqrt{6}$ 7. $18\sqrt{2} - 30$
8. $30 - 105\sqrt{2}$ 9. $9 + 5\sqrt{3}$ 10. $9 + 6\sqrt{5} - 6\sqrt{3} - 4\sqrt{15}$ 11. $-22 - 8\sqrt{15}$ 12. $84 + 6\sqrt{35} - 8\sqrt{21} - 4\sqrt{15}$
13. $17 + 12\sqrt{2}$ 14. $17 - 12\sqrt{2}$ 15. $57 - 12\sqrt{15}$ 16. $141 + 24\sqrt{30}$ 17. 1 18. -29 19. 41 20. 37
21. -1 22. 196 23. 3844 24. $\frac{3}{2}\sqrt{2}$ 25. $\frac{\sqrt{10}}{2}$ 26. $\frac{3}{2}\sqrt{3}$ 27. $\frac{\sqrt{6}}{2}$ 28. $2\sqrt{7}$ 29. $\frac{5}{2}\sqrt{2}$ 30. $\frac{\sqrt{35}}{7}$
31. $\frac{\sqrt{6}}{3}$ 32. $\frac{\sqrt{15}}{6}$ 33. $\frac{\sqrt{14}}{6}$ 34. 2 35. $\sqrt{6}$ 36. $\sqrt{3}$ 37. $\frac{\sqrt{x}}{x^2}$ 38. $\frac{5\sqrt{2x}}{4x}$ 39. $3\sqrt[3]{4}$ 40. $4\sqrt[3]{3}$ 41. $6\sqrt[3]{4}$
42. $\frac{2\sqrt[4]{4x}}{x^2}$ 43. $9\sqrt[4]{2}$ 44. $10\sqrt[4]{3}$ 45. $\frac{7\sqrt[4]{3x^2}}{x}$ 46. $2(3 + \sqrt{5})$ 47. $-12(2\sqrt{2} + 3)$ 48. $\frac{3\sqrt{2} + 2\sqrt{3}}{2}$
49. $3(4\sqrt{5} - 5\sqrt{3})$ 50. $3\sqrt{7} - 2\sqrt{6}$ 51. $-\frac{(49 + 12\sqrt{5})}{41}$ 52. $\frac{-19 + 8\sqrt{3}}{13}$ 53. $\sqrt{h + 4}, 8$ 54. $\frac{1}{\sqrt{h+4}+2}, \frac{1}{4}$
55. $\frac{-1}{2(2+\sqrt{4+h})}, \frac{-1}{8}$ 56. $\sqrt{h + 3}, 6$ 57. $-(2 + \sqrt{4 + h}), -4$ 58. $(x + \sqrt{11})(x - \sqrt{11})$ 59. $(x + 2\sqrt{3})(x - 2\sqrt{3})$
60. $(2x - \sqrt{15})(2x + \sqrt{15})$ 61. $(2\sqrt{3}x - \sqrt{7})(2\sqrt{3}x + \sqrt{7})$ 62. $(2\sqrt{5}x - 3)(2\sqrt{5}x + 3)$
63. $(3\sqrt{2}x - \sqrt{10})(3\sqrt{2}x + \sqrt{10})$ 64. $\pm\sqrt{7}$ 65. $\pm 2\sqrt{3}$ 66. $\frac{\pm\sqrt{6}}{4}$ 67. $\frac{\pm 2\sqrt{6}}{9}$

Exercise set 5

1. $5i$ 2. $6i$ 3. $3\sqrt{2}i$ 4. $3\sqrt{3}i$ 5. -12 6. -6 7. $8\sqrt{2}i$ 8. $5/6$ 9. $-6i$ 10. $-4\sqrt{3}i$ 11. $8 - 5i$ 12. $10 + 6i$
13. $-14 + 33i$ 14. $-15 + 40i$ 15. $21 + i$ 16. $29 + 29i$ 17. $-5 + 12i$ 18. $16 - 30i$ 19. 29 20. 58 21. 22
22. 53 23. $\frac{-15}{2}i$ 24. $\frac{-9}{5}i$ 25. $\frac{-2 - 3i}{4}$ 26. $\frac{7 - 9i}{3}$ 27. $\frac{-5\sqrt{3}i}{6}$ 28. $\frac{-4\sqrt{5}i}{3}$ 29. $2(3 + 2i)$ 30. $-2(2 - 5i)$
31. $2(2\sqrt{3} + 3i)$ 32. $\frac{5 + 12i}{13}$ 33. $\frac{-1 + 43i}{25}$ 34. $\frac{29 + 7i}{20}$ 35. $(x + 2i)(x - 2i)$ 36. $(x + 3i)(x - 3i)$
37. $(x + 2\sqrt{2}i)(x - 2\sqrt{2}i)$ 38. $(2x + 5i)(2x - 5i)$ 39. $(3x + 4i)(3x - 4i)$ 40. $(2\sqrt{3}x + 5i)(2\sqrt{3}x - 5i)$
41. $\pm 5i$ 42. $\frac{\pm 3i}{4}$ 43. $\frac{\pm 2\sqrt{10}}{5}i$ 44. $\frac{\pm 5\sqrt{3}}{6}i$ 46. $-i$ 47. 1 48. -1 49. $-i$ 50. -32 51. 9 52. $\frac{12}{7}i$ 53. $4i$

Exercise 6

1. 2 2. $\frac{1}{2}$ 3. -2 4. $-\frac{1}{2}$ 5. 3 6. -3 7. -3 8. 3 9. $-1/3$ 10. 8 11. $1/8$ 12. 243 13. $1/243$ 14. $-1/5$ 15. $1/4$
16. -4 17. 243 18. 32 19. $1/16$ 20. -8 21. $9/8$ 22. $32/27$ 23. $b^{\frac{11}{8}}$ 24. $b^{\frac{17}{12}}$ 25. $b^{\frac{3}{4}}$ 26. a^4 27. a^2 28. x^2y^3
29. $2x^3y^4$ 30. $9x^6y^2$ 31. $\frac{x^4}{y}$ 32. $\frac{y}{x^4}$ 33. $\frac{8x^6}{27y^9}$ 34. $\frac{4x^4}{9y^8}$ 35. 27 36. 16 37. 125 38. 1 39. 11 40. $31/3$
41. -2 42. 4 43. 0 44. $80/3$ 45. $36/5$ 46. $504/5$ 47. $-209/2$ 48. $\sqrt[4]{125}$ 49. $\sqrt{3}$ 50. $\sqrt[8]{64}$ 51. $\sqrt[15]{256}$
52. $\sqrt[12]{2}$ 53. $\sqrt[12]{3}$ 54. $\sqrt[20]{5}$ 55. $\sqrt[12]{8192}$ 56. $\sqrt[ns]{b^{ms + rn}}$ 57. 8.14% 58. 4.2% 59. 7.05% 60. 3.68%
61. 1.16% 62. 5.79% 63. 8978.60 64. $131,137.78$ 65. 448.65 66 $14,550.23$ 67. $47,073.42$ 68. 7.96%
69. 6.75% 70. 6.46% 71. 7.81% 73(a) 13409.99 (b) 13569.75 (c) 13653.50 (d) 13710.85 (e) 13739.04
(f) 13739.96 74. (a) 13256.06 (b) 13323.74 (c) 13358.70 (d) 13382.45 (e) 13394.08 (f) 13394.45 75. 12yrs
76. 7.95% 77. $m\left[\left(1 + \frac{s}{j}\right)^{\frac{j}{m}} - 1\right]$ 78. (a) 332.99 (b) 333.08 79. (a) 729ft. 6in (b) 2918 ft 1 in 80. (a) 324ft 3 in
(b) 1296 ft 11 in 81. It doubles 82. 36ft. 6 in 83. 1 84 -17 85. $-52/5$ 86. $-14/5, -2$

Exercise set 7

1. $5(2x + 5y)$ 2. $25(5a + 3b)$ 3. $4(3a - 4b)$ 4. $9(2x + 3y)$ 5. $b(a + c)$ 6. $y(b - 1)$ 7. $a(x + 1)$ 8. $24x^2y(5 + 8xy)$
9. $60a^2b^2(a + 2b)$ 10. $30m^3n^2p^3(3p - 5mn)$ 11. $32x^2y^2z^5(3x + 5y^2z)$ 12. $x^{1/2}(4 + 3x)$ 13. $5y^{3/2}(3 + 4y)$
14. $6x^{-1/2}(2 + 3x^2)$ 15. $4x^{-3/2}(5 + 9x^2)$ 16. $2x(x^2 + 1)(3 - x^2)$ 17. $x^4(2x^2 + 3)^6(38x^2 + 15)$

18. $8x^9(2x^2 + x + 7)^{15}(21x^2 + 70)$ 19. $3x^4(x^2 + 2x + 3)^7(5x^2 + 11x + 15)$ 20. $6x^2(x^2 + 1)^{1/2}(3x^2 + 2)$
21. $3x^3(5x^3 + 4)/(x^3 + 1)^{2/3}$ 22. $4x^3(27x^2 + 16)/(3x^2 + 2)^{3/4}$ 23. $(2x + 1)(x - 2)$ 24. $(5x + 2)(2x + 1)$
25. $(3x - 2)(x + 2)$ 26. $(2x + 3)(3x + 2)$ 27. $(2x - 3)(3x + 2)$ 28. $(2x - 3y)(3x - 2y)$ 29. $(4x + 5)(3x - 2)$
30. $(4x - 5y)(3x - 2y)$ 31. $(6x + 5)(3x + 4)$ 32. $(6xy - 5)(3xy + 4)$ 33. $(8xy - 3)(3xy - 8)$ 34. $(8x + 3)(3x - 8)$
35. $4(2x - 3)(3x + 5)$ 36. $3(3x - 2)(2x + 5)$ 37. $-2(2x + 5)(6x - 7)$ 38. $4x(x + 3)(5x - 2)$ 39. $-12x^2(2x - 1)(3x + 2)$
40. $(3x + 4y)^2$ 41. $(x + y)^2$ 42. $(x - y)^2$ 43. $(2x + 3)^2$ 44. $(3xy - 5)^2$ 45. $(5x - 3)^2$ 46. $(x - 2)(x + 2)$
47. $(2x - 3)(2x + 3)$ 48. $(8x - 5y)(8x + 5y)$ 49. $(4x - 7)(4x + 7)$ 50. 3, 4   51. 2, 6   52. -2/3, 3/2   53. -5/4, 2/3
54. -3, 0, 2/5   55. -2/3, 0, ½   56. 0, 7/3   57. 0, 3/4   58. -5/3, 5/3   59. -2, 2/3   60. -½, 2   61. $(x + 3)(w + 2)$
62. $(3a + 2)(2 + b)$ 63. $(y + 2)(x + 3)$ 64. $(ab + 2)(bc + 3)$ 65. $(2x + 1)(x - 2)$ 66. $(3x - 2)(x + 2)$
67. $(2x + 3)(3x + 2)$ 68. $(4x + 5)(3x - 2)$ 69. $(3x + 4)^2$ 70. $(5x - 3)^2$

Exercise set 8

1. 16   2. -6   3. 0   4. 0   5. 8/3   6. 4   7. -2/9   8. 26/11   9. 59/21   10. 2   11. -3/7   12. -14/11   13. 12   14. 15
15. 8/15   16. 10   17. -33   18. 29/7   19. 19/18   20. 65/84   21. 755/9   22. $(19 + 2y)/5$ 23. $(5x - 19)/2$
24. $(c - by)/a$ 25. $(c - ax)/b$ 26. $5/9(F - 32)$ 27. $V/(\pi r^2)$ 28. $(2A - ah)/h$ 29. $RW/(W-R)$ 30. $M = 19, L = 12$
31. $J = 16, M = 21$ 32. 84   33. $w = 5, l = 13$ 34. 12 cm   35. {31,32,33}   36. {17,19,21}   37. (a) 2 hrs
(b) 160 mi, 120 mi   38. 4 hrs   39. 4 hrs   40. 25°   41. 30°   42. 36   43. 5/14   44. 12/15

Exercise set 9

1. ±5   2. ±6   3. ±2   4. ±7   5. ±8   6. ±3   7. ±5   8. ±8   9. ±3   10. ±3   11. $\pm 3\sqrt{2}$ 12. $\pm 2\sqrt{2}$ 13. $\pm 2\sqrt{3}$
14. $\pm 2\sqrt{6}$ 15. ±5i   16. ±6i   17. ±2i   18. ±7i   19. ±3i   20. ±4i   21. $\pm 2\sqrt{5}i$ 22. $\pm 2\sqrt{2}i$ 23. ±4i
24. $\pm 3\sqrt{2}i$ 25. $\pm 3\sqrt{3}i$ 26. $\pm 2\sqrt{3}$ 27. $\pm 4\sqrt{2}i$ 28. $\frac{\pm\sqrt{170}}{10}$ 29. $\frac{\pm\sqrt{230}}{10}$ 30. $\pm 2\sqrt{6}i$ 31. -1, 5   32. -4, 5
33. -5/2, 3   34. -8,2   35. -2, 10   36. -2, 5   37. -4/3, 4   38. $\frac{7\pm3\sqrt{2}}{4}$ 39. $\frac{-4\pm2\sqrt{5}}{5}$ 40. $\frac{3\pm2\sqrt{6}}{2}$ 41. $\frac{1\pm2\sqrt{2}}{3}$
42. $-3 \pm 3i$ 43. $4 \pm 4i$ 44. $5\pm2\sqrt{3}i$ 45. $\frac{2\pm5i}{3}$ 46. $\frac{-7\pm6i}{4}$ 47. $\frac{3\pm2\sqrt{6}i}{2}$ 48. $\frac{5\pm3\sqrt{2}i}{3}$ 49. $\frac{2\pm2\sqrt{3}i}{3}$ 50. $-3\pm3\sqrt{2}$
51. $\frac{2\pm2\sqrt{3}}{3}$ 52. $\frac{-4\pm5\sqrt{5}i}{10}$ 53. $\frac{3\pm2\sqrt{5}i}{2}$ 54. $-7\pm3\sqrt{2}$ 55. $\frac{-7\pm3\sqrt{2}i}{2}$ 56. $\frac{15\pm2\sqrt{15}}{9}$ 57. $\frac{15\pm2\sqrt{15}i}{6}$

Exercise set 10

1. $(x + 2)^2 - 4$ 2. $(x + 3)^2 - 9$ 3. $(x - 3)^2 - 9$ 4. $(x - 4)^2 - 16$ 5. $(x - 5)^2 - 25$ 6. $(x + 6)^2 - 36$ 7. $(x + 3/2)^2 - 9/4$
8. $(x - 3/2)^2 - 9/4$ 9. $(x - 5/2)^2 - 25/4$ 10. $(x + 5/2)^2 - 25/4$ 11. $(x - 7/2)^2 - 49/4$ 12. $(x + 9/2)^2 - 81/4$
13. $-1\pm\sqrt{11}$, $-4.317, 2.317$ 14. $-3\pm\sqrt{6}$, $-0.551, -5.449$ 15. $1\pm\sqrt{11}$, $-2.317, 4.317$
16. $3\pm\sqrt{6}$, $0.551, 5.449$ 17. $-3\pm2\sqrt{3}$, $-6.464, 0.464$ 18. $4\pm2\sqrt{5}$, $-0.472, 8.472$
19. $6\pm2\sqrt{10}$, $-0.325, 12.325$ 20. $\frac{-5\pm3\sqrt{5}}{2}$, $-5.854, 0.854$ 21. $\frac{5\pm3\sqrt{5}}{2}$, $-0.854, 5.854$
22. $\frac{3\pm\sqrt{13}}{2}$, $-0.303, 3.303$ 23. $\frac{-3\pm\sqrt{13}}{2}$, $-3.303, 0.303$ 24. $-1 \pm 3i$ 25. -2, 5   26. -7, 5   27. 1, 1/6
28. $-1\pm2\sqrt{7}$, $-6.29, 4.29$ 29. $\frac{-5\pm\sqrt{39}i}{2}$ 30. $\frac{-1\pm\sqrt{119}i}{12}$ 31. $3\pm\sqrt{14}i$ 32. $6\pm2\sqrt{3}i$ 33. $-5 \pm 5i$ 34. $4\pm2\sqrt{5}i$
35. $2\pm2\sqrt{3}$, $-1.464, 5.464$ 36. $-3\pm2\sqrt{3}i$ 37. $\frac{3\pm\sqrt{3}i}{2}$ 38. $\frac{-5\pm\sqrt{15}i}{2}$ 39. $\frac{-4\pm\sqrt{14}}{2}$, $-3.871, -0.129$ 40. -2, 3/2
41. -3/2, 2/3   42. -3/4, 4/3   43. -5/2, -4/3   44. $\frac{-5\pm\sqrt{55}i}{4}$ 45. $\frac{-5\pm\sqrt{103}i}{8}$ 46. $-5\pm\sqrt{21}$, $-9.583, -0.417$
47. $\frac{9\pm\sqrt{321}}{12}$, $-0.743, 2.243$ 48. $\frac{15\pm\sqrt{393}}{12}$, $-0.402, 2.902$ 49. $\frac{-2\pm\sqrt{7}}{2}$, $-2.323, 0.323$
50. $\frac{5\pm3\sqrt{5}}{10}$, $-0.171, 1.171$ 51. $b = 8, h = 5$ 52. 2.34 sec   53. 16.2 m by 6.2 m

Exercise set 11

Answers in 1-15 are in the form (a, b, c)
1. (1,4,-11)   2. (3,7,-8)   3. (5, -8, -12)   4. (3,7,-10)   5. (5,27,-10)   6. (9,-4,4)   7. (5,10,-14)   8. (2,0,-10)
9. (5,0,11)   10. (7,-3,0)   11. (4,-9,0)   12. (3,-2,-13)   13. (8,-14,-22)   14. (60,-152,71)   15. (4,5,-21)

16. $-1\pm\sqrt{11}$, -4.313, 2.317 17. $-3\pm\sqrt{6}$, -0.551, -5.449 18. $1\pm\sqrt{11}$, -2.317, 4.317

19. $3\pm\sqrt{6}$, 0.551, 5.449 20. $-3\pm2\sqrt{3}$, -6.464, 0.464 21. $4\pm2\sqrt{5}$, -0.472, 8.472

22. $6\pm2\sqrt{10}$, -0.325, 12.325 23. $\frac{-5\pm3\sqrt{5}}{2}$, -5.854, 0.854 24. $\frac{5\pm3\sqrt{5}}{2}$, -0.854, 5.854

25. $\frac{3\pm\sqrt{13}}{2}$, -0.303, 3.303 26. $-1\pm3i$ 27. -2, 5 28. -7, 5 29. 1, 1/6 30. $-1\pm2\sqrt{7}$, -6.29, 4.29

31. $\frac{-5\pm\sqrt{39}i}{2}$ 32. $\frac{-1\pm\sqrt{119}i}{12}$ 33. $3\pm\sqrt{14}i$ 34. $6\pm2\sqrt{3}i$ 35. $-5\pm5i$ 36. $4\pm2\sqrt{5}i$ 37. $2\pm2\sqrt{3}$, -1.464, 5.464

38. $-3\pm2\sqrt{3}i$ 39. $\frac{3\pm\sqrt{3}i}{2}$ 40. $\frac{-5\pm\sqrt{15}i}{2}$ 41. $\frac{-4\pm\sqrt{14}}{2}$, -3.871, -0.129 42. -2, 3/2 43. -3/2, 2/3 44. -3/4, 4/3

45. -5/2, -4/3 46. $\frac{-5\pm\sqrt{55}i}{4}$ 47. $\frac{-5\pm\sqrt{103}i}{8}$ 48. $-5\pm\sqrt{21}$, -9.583, -0.417 49. $\frac{9\pm\sqrt{321}}{12}$, -0.743, 2.243

50. $\frac{15\pm\sqrt{393}}{12}$, -0.402, 2.902 51. $\frac{-2\pm\sqrt{7}}{2}$, -2.323, 0.323 52. $\frac{5\pm3\sqrt{5}}{10}$, -0.171, 1.171 53. 5, 6, 7, or -7, -6, -5

54. 11, 13, 15, or -15, -13, -11 55. 16.971 in 56. 13.416 in 57. (a) 0.195 sec or 4.805 sec (b) 5 sec (c) 5.807 sec

58. (a) 3.062 sec (b) 3.536 sec (c) 3.953 sec 59. 4ft 60. 82 and 94 miles 61. (a) 3.584ft (b) 12.832ft by

16.832 ft 62. 10 and 12 mph 63. 4 hr 64. -4/3, 8 65. -17/12, 3 66. 6/13, 4 67. ±2 68. $\frac{\sqrt[3]{12}}{2}$, $\frac{\sqrt[3]{36}}{3}$

69. 512, 19683 70. $\pm\sqrt{\frac{2+\sqrt{10}}{2}}$

Exercise set 12.

Note the following abbreviations: RDR = real (double)repeated root, RRR = Real Rational Roots, RIR = Real Irrational Roots, RCR = Real Complex Roots. 1. (a) RDR (b) $x = 9$ 2. (a) RDR (b) $x = 1$ 3. (a) RDR (b) $x = 4$
4. (a) RDR (b) $x = 3$. 5. (a) RDR (b) $x = -2$ 6. (a) RDR (b) $x = -5$ 7. (a) RDR (b) x = -3 8.(a) RRR (b) $x = 3/2, -4/3$
9. (a) RRR (b) $x = -3/2, 1/5$ 10. (a) RRR (b) $x = -4/3, 3/4$ 11. (a) RRR (b) $x = -5/2, 3/7$ 12. (a) RRR (b) $x = -3$,
-1/2 13. (a) RRR (b) $x = -5/2, 7/6$ 14. (a) RRR (b) $x = 1$. 10/3 15. (a) RIR (b) $x = \frac{4\pm\sqrt{10}}{3}$ 16. (a) RIR
(b) $x = \frac{-5\pm\sqrt{17}}{4}$ 17. (a) RIR (b) $x = \frac{-1\pm\sqrt{22}}{2}$ 18. (a) RIR (b) $x = \frac{-4\pm\sqrt{26}}{5}$ 19 (a) RIR (b) $x = \frac{5\pm3\sqrt{5}}{2}$ 20 (a) RIR
(b) $x = \frac{15\pm\sqrt{2145}}{40}$ 21 (a) RCR (b) $x = 1\pm\sqrt{2}i$ 22. (a) RRR (B) $x = -1/2, 2$ 23. (a) RCR $x = \frac{-2\pm\sqrt{2}i}{3}$

24. (a) RCR (b) $x = \frac{2\pm\sqrt{6}i}{5}$ 25. (a) RCR (b) $x = \frac{-5\pm\sqrt{71}i}{16}$ 26. $x^2 - 8x + 15 = 0$. 27. $x^2 - 11x + 28 = 0$

28. $x^2 - 3x - 10 = 0$ 29. $x^2 + x - 42 = 0$ 30. $x^2 - 6x + 9 = 0$ 31. $x^2 - 10x + 25 = 0$ 32. $x^2 + 6x + 9 = 0$
33. $6x^2 - 19x + 15 = 0$ 34. $12x^2 + 11x + 2 = 0$ 35. $35x^2 + 8x - 3 = 0$ 36. $x^2 - 2x - 1 = 0$ 37. $x^2 - 4x - 41 = 0$
38. $x^2 - 10x + 1 = 0$ 39. $x^2 - 6x - 71 = 0$ 40. $x^2 - 6x + 10 = 0$ 41. $x^2 - 2x + 5 = 0$ 42. $x^2 - 10x + 34 = 0$
43. $x^2 - 2x + 3 = 0$ 44. $x^2 - 4x + 49 = 0$ 45. $x^2 - 10x + 49 = 0$ 46. $x^2 - 6x + 89 = 0$ 47. (a) 18 (b) 81 48. (a) 2
(b) 1 49. (a) 8 (b) 16 50. (a) 6 (b) 9 51. (a) -4 (b) 4 52. (a) -10 (b) 25 53. (a) -6 (b) 9 54. (a) 1/6 (b) -2
55. (a) -13/10 (b) -3/10 56, (a) -7/12 (b) -1 57. (a) -29/14 (b) -15/14 58. (a) -7/2 (b) 3/2 59. (a) -4/3 (b) -35/12
60. (a) 13/3 (b) 10/3 61. (a) 8/3 (b) 2/3 62. (a) -5/2 (b) 1/2 63. (a) -2/3 (b) -7/3 64. (a) -8/5 (b) -2/5 65. (a) 5
(b) -5 66. (a) 3/4 (b) -6/5 67. (a) 2 (b) 3 68. (a) 3/2 (b) -1 69. (a) -4/3 (b) 2/3 70. (a) 4/5 (b) 2/5 71. (a) -5/8
(b) 3/8 73. (a) Let the roots be x_1, x_2, x_3 then we have $x_1 + x_2 + x_3 = -b/a$, $x_1 x_2 + x_1 x_3 + x_2 x_3 = c/a$, $x_1 x_2 x_3 = -d/a$
(b) the sum of the roots = $-a_{n-1}/a_n$, the sum of the roots multiplied together two at a time = a_{n-2}/a_n, the sum of the
roots multiplied together three at a time = $-a_{n-3}/a_n$, ... and the product of the roots = $(1)^n a_0/a_n$ 74. $c_2 = 0$ and $b^2 - 4ac_1 \geq 0$

Exercise set 13

1. 25 2. 9 3. 35 4. 8 5. 49/5 6. 4 7. 6 8. 2 9. 3 10. 8 11. 4 12. -2 13. 5/3, 2 14. 5 15. 4 16. 6
17. 6 18. 3 19. 1 20. 1 21. 3 22. 4 23. 5 24. 3. 25. ±1 26. illegal squaring 27. square root is positive
28. 0, 1/4 29. 1 30. 25 31. 36 32. ±6 33. 2500 watts 34. 6.96 miles/sec

Exercise set 14

1. $x > 3/2$ 2. $x \leq 3/2$ 3. $x \leq -4$ or $x \geq 3$ 4. $-4 < x < 3$ 5. $-5/3 \leq x \leq 7/4$ 6. $x < -5/3$ or $x > 7/4$ 7. $x \leq -5$ or $x \geq 3/5$ 8. $-5 < x < 3/5$ 9. $-4 < x < 5/2$ 10. $x \leq 2/3$ or $x \geq 3/4$ 11. $x \leq 0$ or $x \geq 5/2$ 12. $0 < x < 5/2$ 13. $x = -2$ or $x \geq 3$ 14. $x \leq 3/2$ 15. $x \leq -3$ or $0 \leq x \leq 2$ 16. $-5 \leq x \leq -1$ or $x \geq 2$ 17. $x \leq -3/2$ or $-1 \leq x \leq 5/3$ 18. $x \leq -4$ or $x \geq -3$ 19. $2 < x < 5$ 20. $x < -7$ or $x > 6$ 21. $-4/3 \leq x \leq 3/2$ 22. $x < -4/3$ or $x > 3/2$ 23. $x \leq -3/2$ or $x \geq 4/5$ 24. $-3/2 < x < 4/5$ 25. $x \leq -5/4$ or $x \geq 5/6$ 26. $-5/4 < x < 5/6$ 27. $-4 \leq x \leq 0$ or $x \geq 4$ 28. $x < -4$ or $0 < x < 4$ 29. $-5 \leq x \leq 0$ or $x \geq 5$ 30. $x < -5$ or $0 < x < 5$ 31. $-4 \leq x \leq 4$ 32. $x < -1$ or $x > 1$ 33. $x \leq -5/2$ or $x \geq 4$ 34. $-5/2 < x < 4$ 35. $-17/6 < x < 2$ 36. $x \leq -17/6$ or $x \geq 2$ 37. $x \leq -11/20$ or $x \geq 1$ 38. $-11/20 < x < 1$ 39. $x \leq -2$ or $1 \leq x < 3$ 40. $-2 < x < 1$ or $x > 3$ 41. $x < 2/3$ or $5/2 < x < 4$ or $x > 4$ 42. $2/3 < x < 5/2$ 43. $x \leq 0$ or $x = 3/2$ or $x > 4$ 44. $0 < x < 3/2$ or $3/2 < x < 4$ 45. $x < 3$ or $x > 8$ 46. $3 \leq x \leq 8$ 47. $x < -3$ or $x \geq 4$ 48. $-3 < x < 4$ 49. $1 < x \leq 3$ 50. $x < 1$ or $x > 3$ 51. $x > 3$ 52. $x < 3$ 53. $x \leq -6$ or $-5 < x < 5$ or $x \geq 6$ 54. $-6 < x < -5$ or $5 < x < 6$ 55. $x \leq -3$ or $x = 0$ or $\frac{1}{2} < x \leq 2$ 56. $-3 < x < -4/3$ or $-4/3 < x < 0$ or $0 < x < \frac{1}{2}$ or $x > 2$ 57. $x < -9/2$ or $7/4 \leq x < 2$ or $x = 0$ 58. $-9/2 < x < -1/2$ or $-1/2 < x < 0$ or $0 < x < 7/4$ or $x > 2$ 59. $-2\sqrt{2} < x < 2\sqrt{2}$ 60. $x \leq -2\sqrt{2}$ or $x \geq 2\sqrt{2}$ 61. $1 - \sqrt{2} < x < 1 + \sqrt{2}$ 62. $x \leq 1 - \sqrt{2}$ or $x \geq 1 + \sqrt{2}$ 63. $x < 2 - 3\sqrt{5}$ or $x > 2 + 3\sqrt{5}$ 64. $2 - 3\sqrt{5} < x < 2 + 3\sqrt{5}$ 65. $\emptyset$ 66. $-\infty < x < \infty$

Exercise set 15

1. $y = 2x + 4$ 2. $y = -3x$ 3. $y = -2$ 4. $y = \frac{1}{2}x + 1/4$ 5. $y = \frac{1}{4}x + 3$ 6. $3/4$, $(4,0)$, $(0,-3)$ 7. $-\frac{1}{2}$, $(9/2,0)$, $(0,9/4)$ 8. $5/4$, $(-12/5, 0)$, $(0,3)$ 9. $-13/47$, $(-112/13, 0)$, $(0,-112/47)$ 10. 0, $(0,-19/4)$ 11. $y = -3x + 11$ 12. $y = \frac{1}{2}x + 1$ 13. $y = \frac{1}{6}x$ 14. $y = 14x + 143/10$ 15. $y = 4.1x - 12.3$ 16. $x = -1$ 17. $y = -6$ 18. $y = 2x + 4$ 19. $x = 12$ 20. $y = 8$

21.

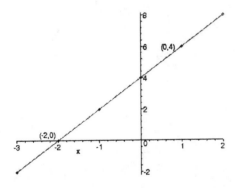

22.

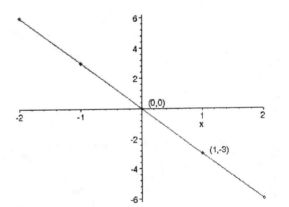

23.

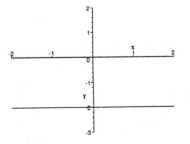

24.

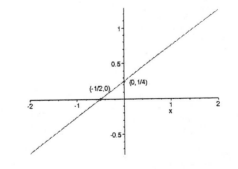

25.

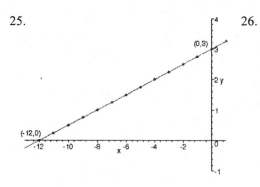

26.

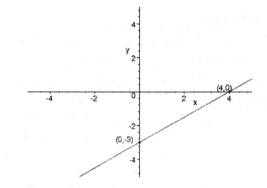

27.

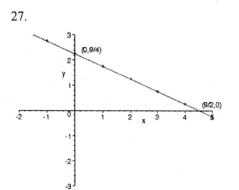

28.

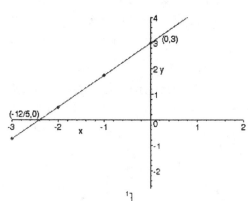

29.

30.

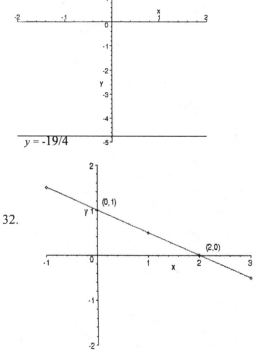

$y = -19/4$

31.

32.

33.

34.

35.

36.

$x = -1$

37.

38.

39.

$x = 12$

40.

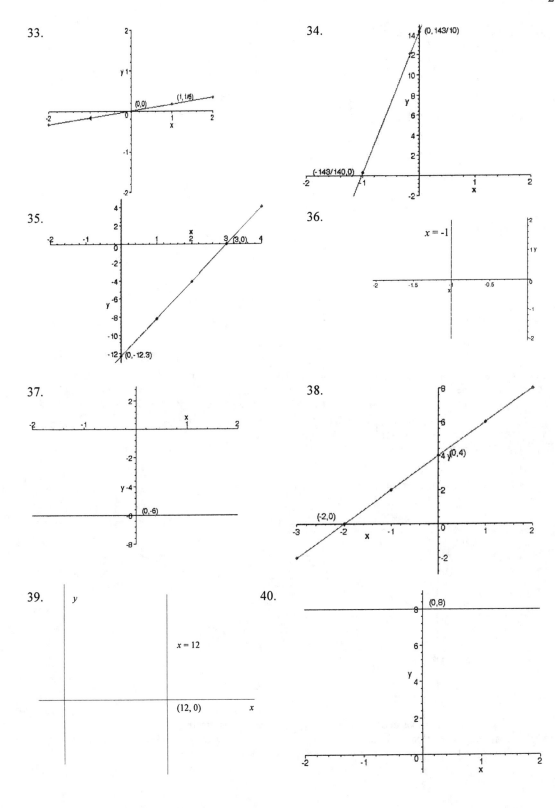

216

41. (a) 2/5, (3,0), (0,-6/5) (b) $y = \frac{2}{5}x + \frac{37}{5}$

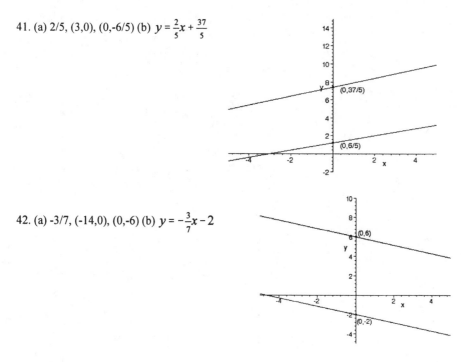

42. (a) -3/7, (-14,0), (0,-6) (b) $y = -\frac{3}{7}x - 2$

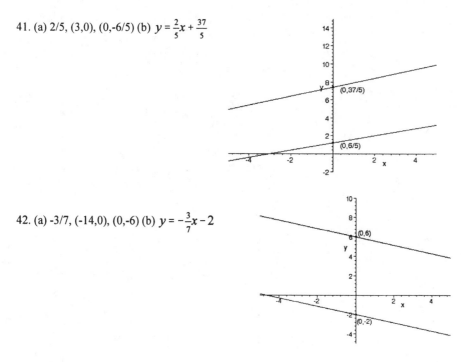

43. (a) $y = -6/5$ (b) $x = 3$ 44. (a) $y = -6$ (b) $x = -14$ 45. $y = 1, y = -7$ 46. $x = 8, x = -4$ 47. no slope   48. 7/5   49. -8/5   50. 0   51. no slope   52. ½   53. -7   54. 24   55. -2   56. 0   57. no slope   58. 4/7   59. $x = 1$ 60. $7x - 5y = 59$
61. $8x + 5y = 0$.   62. $y = 4$ 63. $x = 12$ 64. $y = ½x + 3$ 65. $14x + 2y = 3$ 66. $y = 24x - 9$ 67. $y = -2x$ 68. $y = -5$
69. $x = ½$ 70. $y = \frac{4}{7}x + 4$ 71. $y = 3x - 8$ 72. (a) $2x - 5y = 16$ (b) $5x + 2y = -18$ 73. (a) $3x + 7y = -18$

74. (a) 3 (b) $7x - 3y = 16$ 74 .(b) $\dfrac{C^2}{2|AB|}$ 75. $5y + 2x = 20, 5y - 2x = 20$.   76. $y = -5x \pm 20$ 77. (a) 91/5

(b) $\left|\dfrac{C}{AB}\right|\sqrt{A^2 + B^2}$ 78.(b) no (c) examine the slopes   79. $t = -0.05x + 4000$ 80. 130   81. 166

82.

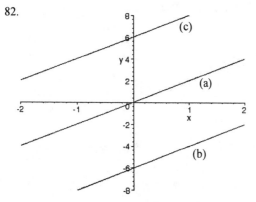

83.

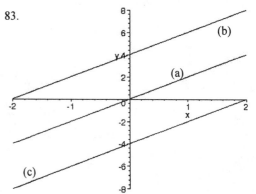

(d) They are horizontal translations of each other. Each graph may be obtained from (a) by moving it 3 units to the left or right.

(d) They are vertical translations of each other. Each graph may be obtained from (a) by moving it three units Up or down.

84.

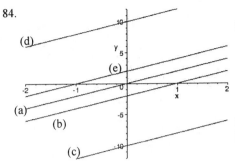

(e) They are each obtained from (a) by moving it 4 units up or down and simultaneously moving it 3 units to the left or right.

85. The second line is obtained from the first by moving it *h* units horizontally and *k* units vertically.

Exercise set 16

1. C(3,2) r = 2

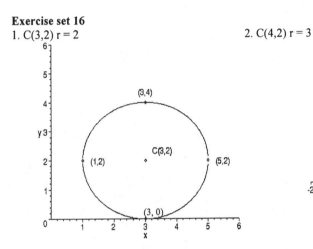

2. C(4,2) r = 3

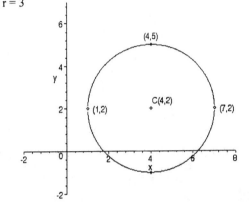

3. C(-4,3) r =4

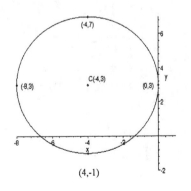

4. C(3,-2) r = 5

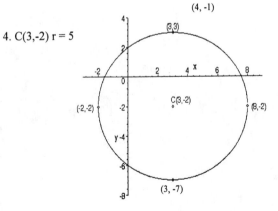

218

5. C(-1,-4) r =4

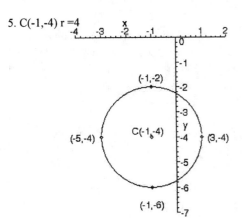

6. C(-3,-2) r = 3

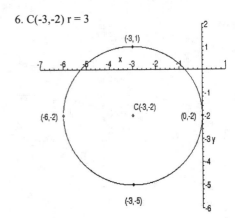

7. C (0,0) r = 2

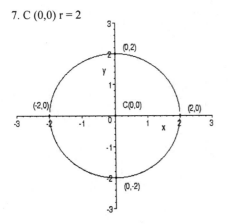

8. C(0,0) r = 5

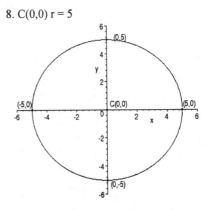

9. C(-3,5) r = 5

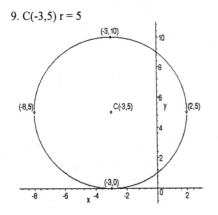

10. C(1,-2) r = 2

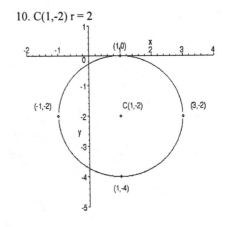

11. C(-4,3) r = 4

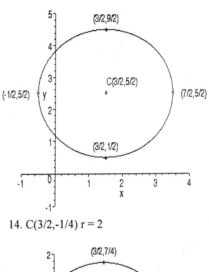

12. C(3/2,5/2) r = 2

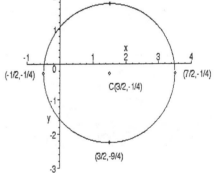

13. C(2/3,-4/3) r = 11/3

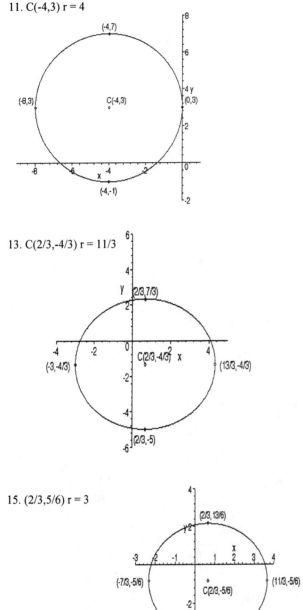

14. C(3/2,-1/4) r = 2

15. (2/3,5/6) r = 3

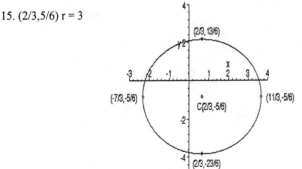

16. Contradiction   17. Contradiction   18. Contradiction   19. Circle   20. Point   21. Point   22. Contradiction
23. point   24. Point 25. Circle   26 (a) $b^2 + c^2 -4ad > 0$ (b) $b^2 + c^2 -4ad = 0$ (c) $a = 0$, $bc \neq 0$,
(d) $b^2 + c^2 -4ad < 0$ 27. (a) $2\sqrt{3}$ (b) 4 (c) $2\sqrt{3}$ 28. (a) $-2\sqrt{3}$ (b) -4 (c) $-2\sqrt{3}$ 29. (a) $3 + \sqrt{15}$ (b) $3 + 2\sqrt{5}$
(c) $3 + \sqrt{7}$ 30. (a) $3 - \sqrt{15}$ (b) $3 - 2\sqrt{5}$ (c) $3 - \sqrt{7}$ 31. (a) $5 + 2\sqrt{6}$ (b) $5 - 2\sqrt{6}$ (c) 8 (d) 2

220

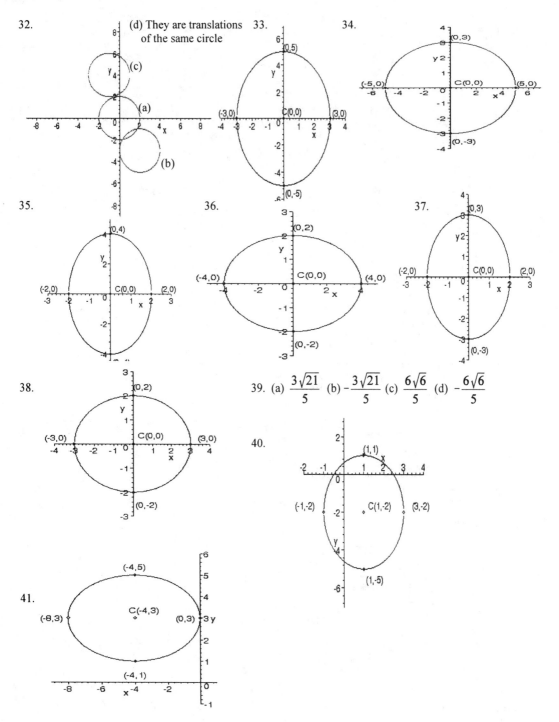

32.

(d) They are translations of the same circle

33.

34.

35.

36.

37.

38.

39. (a) $\dfrac{3\sqrt{21}}{5}$ (b) $-\dfrac{3\sqrt{21}}{5}$ (c) $\dfrac{6\sqrt{6}}{5}$ (d) $-\dfrac{6\sqrt{6}}{5}$

40.

41.

Exercise set 17

1. (2, 3) 2. (-2, 3) 3. (4, -5) 4. (5, 2) 5. (½, 1/3) 6. (3/4, -1/5) 7. (3/5, 5/7) 8. (3/8, -2/3) 9. (8, -10)
10. (30, 12) 11. (1/3, ½) 12. (-1/3, ½) 13. (8, 13) 14. (-2, -1) 15. (-3, 4) 16. (2, -3) 17. (7, 9) 18. (5, 4) 19. (-2, 3) 20. (½, 1/3)

Exercise set 18

Answers in the form (x, y) or (x, y, λ) or (x, y, z, λ), whichever is appropriate

1. (-1, -6), (2, 3) 2. (-1,4), (2, -2) 3. $(-2\sqrt{2}, -\sqrt{2}), (2\sqrt{2}, \sqrt{2})$ 4. $(-\sqrt{3}, -2\sqrt{3}), (\sqrt{3}, 2\sqrt{3})$ 5. (5/3, -11/3), (3, -1)
6. (56/17, -15/17), (2, 3) 7. (1, 3), (9/19, 109/38) 8. (3, -2), (-179/63, -134/21) 9. (-3,2), (509/92, -65/46)
10. $(-\sqrt{2}, -\sqrt{2}), (-\sqrt{2}, \sqrt{2}), (\sqrt{2}, -\sqrt{2}), (\sqrt{2}, \sqrt{2})$, 11. $(2\sqrt{3}, \pm 2\sqrt{2}), (-2\sqrt{3}, \pm 2\sqrt{2})$
12. $(-3\sqrt{2}, -3\sqrt{3}), (-3\sqrt{2}, 3\sqrt{3}), (3\sqrt{2}, -3\sqrt{3}), (3\sqrt{2}, 3\sqrt{3})$ 13. (-3, -2), (-3, 2), (3, -2), (3, 2)
14. (-4, -3), (-4, 3), (4, -3), (4, 3) 15. (-2, 3), (2, 3), $(\frac{-3\sqrt{2}}{2}, -2\sqrt{2}), (\frac{3\sqrt{2}}{2}, 2\sqrt{2})$

16. (-1, -½), (1, ½), $(-\frac{\sqrt{3}}{3}, -\frac{\sqrt{3}}{2}), (\frac{\sqrt{3}}{3}, \frac{\sqrt{3}}{2})$ 17. (-2, -6/5), (2, 6/5), (-3, -4/5), (3, 4/5) 18. (-2, -1/5), (2, 1/5)
19. (1, 5), (2, 8) 20. (3, 5), (22, -9/2) 21. (-3, 4), (3, 4), (0, -5) 22. (-5, 12), (5, 12), (0, -13)
23. (-143/29, 24/29), (3, 4) 24. (3,1), (-71/37, -93/37) 25. (-2, -1), (2, 1), $(-\sqrt{2}, -\sqrt{2}), (\sqrt{2}, \sqrt{2})$

26. (-3, -1), (3,1), $(-\frac{\sqrt{15}}{3}, -\frac{3\sqrt{15}}{5}), (\frac{\sqrt{15}}{3}, \frac{3\sqrt{15}}{5})$ 27. (-4, -2), (4, 2) 28. (-2, 3), (2, -3),
$(\frac{-3\sqrt{15}}{5}, \frac{2\sqrt{15}}{3}), (\frac{3\sqrt{15}}{5}, \frac{-2\sqrt{15}}{3})$ 29.(30,20,50) 30. (15, 45, 15/2) 31. (8, 4, 16/5, 32/5) 32. (10, 30, 6, 30)
33. (2, 8, 10, 8/5) 34. (2, $2\sqrt{3}$, 2, 16) 35. (1, 2, 2, 2) 36. (12, 3, 2, 10368)

37. (a)   (b)   (c)

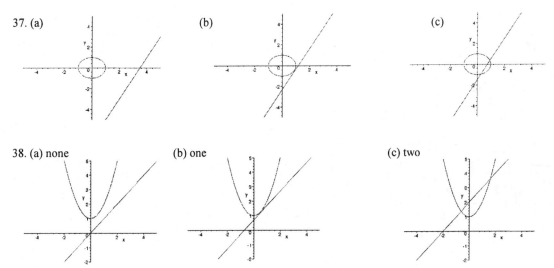

38. (a) none   (b) one   (c) two

222

39. (a) no intersections   (b) one   (c) two   (d) three

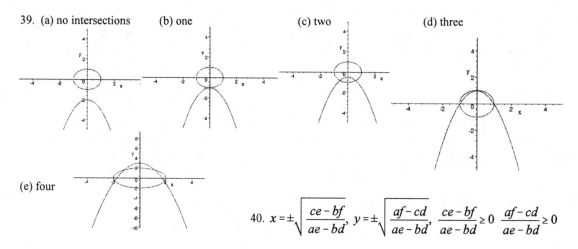

(e) four

40. $x = \pm\sqrt{\dfrac{ce-bf}{ae-bd}}, \quad y = \pm\sqrt{\dfrac{af-cd}{ae-bd}}, \quad \dfrac{ce-bf}{ae-bd} \geq 0 \quad \dfrac{af-cd}{ae-bd} \geq 0$

Exercise set 19

1. -17   2. -5   3. -24   4. 34   5. -1   6. -21   7. 0   8. -59   9. -44   10. 168   11. 68   12. 10   13. -42   14. 180
15. -300   16. 50   17. -68   18. 0   19. -5   20. 5   21. -6   22. -6   23. 24   24. 9   25. -147   26. 11   27. 66
28. 14   29. -79   30. -19

22. INDEX

Abscissa, 142
Absolute Value, 199
Addition Property, 77
Addition-Subtraction Method. 175
Additive Inverse, 6
Algebraic Expressions, 2
Associative Property, 3
Base, 12
Center of Circle, 161
Circle, 159
Coefficients, 2
Commutative Property, 4
Completing the Square, 95
Complex Number, 46
Complex Conjugate, 4
Compound Interest, 54
Conjugates, 35
Coordinate System, 142
Cross Multiplication, 85
Denominator, 2
Difference of Two Squares, 70
Discriminant, 113
Distance Formula, 26-27
Distributive Property, 4
Double (Repeated) Root, 114
Ellipse, 166
Empty Set, 135
Exponents, 2, 11, 15
Extraneous Roots, 121
Factoring, 61-73
Factoring by Grouping, 71
Factors, 2
FOIL, 35, 65
Fractional Root, 51
GCF, 61
General Form of Circle, 162
General Linear Equation, 151
Horizontal Line, 143
Hypothenuse, 23
i, 45
Identity, 6
Imaginary Number, 46
Indeterminate Forms, 39
Integers, 1
Interval Notation, 131
Irrational Numbers, 2, 20

Lagrange Multiplier, 191
LCD, 79
Linear Equations, 77-85
Method of Elimination, 173, 187
Method of Substitution, 176, 182
Monomial Factors, 64
Multiplication Property, 77
Multiplicative Inverse, 6
Natural Numbers, 1
Non-Linear Systems of Equations, 181
Numerator, 2
Nth Root, 20
Order of Operations, 2
Ordinate, 142
Origin, 142
Parallel Lines, 144
Perfect Square, 70
Perpendicular Lines, 154
Pi (π), 20, 56
Point-Slope Equation, 147
Positive integers, 1
Product of Roots, 116
Pythagoras's Theorem, 23
Quadratic Formula, 104
Radical, 20
Radicand, 20
Radius of Circle, 159
Rate, 84
Rational Numbers, 2, 20
Rationalizing, 36
Right Triangle, 23
Rule of 72, 55
Sign Analysis, 129-136
Simple Pendulum, 56
Slope Intercept Form, 145
Slope, 143, 149
Solving Quadratic Equations by Factoring, 72
Square Root, 19
Standard Form of Circle, 162
Sum of Roots, 116
Terms, 2
Transposition, 77
Variables, 2
Vertical Line, 143
X-intercept 152
Y-intercept, 144